Computer Numerical Control Programming

Computer Numerical Control Programming

Michael Sava

Joseph Pusztai

Humber College

Prentice Hall, Englewood Cliffs, New Jersey 07632

Library of Congress Cataloging-in-Publication Data

Sava, Michael.
   Computer numerical control programming / Michael Sava and Joseph
Pusztai.
     p.   cm.
   Includes index.
   1. Machine-tools—Numerical control—Programming.   I. Pusztai,
Joseph.   II. Title.
TJ1189.S29   1990
621.9′023—dc20
                                         89-33058
                                         CIP

Editorial/production supervision: Karen Winget
Cover design: 20-20 Services, Inc.
Manufacturing buyer: Paula Massenaro

© 1990 by Prentice-Hall, Inc.
A Division of Simon & Schuster
Englewood Cliffs, New Jersey 07632

Printed in the United States of America
10 9 8 7 6 5 4 3 2 1

0-13-156084-0

PRENTICE-HALL INTERNATIONAL (UK) LIMITED, *London*
PRENTICE-HALL OF AUSTRALIA PTY. LIMITED, *Sydney*
PRENTICE-HALL CANADA INC., *Toronto*
PRENTICE-HALL HISPANOAMERICANA, S.A., *Mexico*
PRENTICE-HALL OF INDIA PRIVATE LIMITED, *New Delhi*
PRENTICE-HALL OF JAPAN, INC., *Tokyo*
SIMON & SCHUSTER ASIA PTE. LTD., *Singapore*
EDITORA PRENTICE-HALL DO BRASIL, LTDA., *Rio de Janeiro*

Contents

Chapter 5
CUTTER CENTERLINE PROGRAMMING 71

Chapter 6
TOOL OFFSETS 93

Chapter 7
CUTTER DIAMETER COMPENSATION AND TNR COMPENSATION: PROGRAMMING THE WORK SURFACE 101

Contents

Chapter 12
COMPUTERIZED CNC   210

APPENDIX A

INDEX   297

Preface

Islands of automation have brought about more changes in manufacturing in the last 5 years than in the previous 40 years. Microprocessors and computers now assist and direct more than 80 percent of our manufacturing processes, from design, through production engineering, to manufacturing and sales.

The principles of CNC technology have been extended to the manufacturing industry at large from manufacturing cells and flexible manufacturing systems to the total concept of computer integrated flexible manufacturing (CIFM). Our present-day manufacturing systems are integrated with automatically guided vehicles (AGVs) and robotics, which can also be programmed off-line using standard, CNC-like languages.

The people who work in this environment face a continuous process of updating as the new technology unfolds. Learning must be regarded as a long-term investment. It is a proven fact that there is a substantial cost associated with lack of training, in poor quality, costly accidents, low morale, and unacceptably low productivity. The greatest challenge faced by educators is training for a manufacturing environment metamorphosed by a revolutionary technology.

This text covers concepts and fundamentals, manual programming, offsets, compensation, canned cycles, and other standard features. In addition, it carries out extensive coverage of the very latest computer-aided programming. The software described is used extensively by industry and educational institutions. The high potential of user macros is explored as well through detailed programs and explanations, and the reader will get a better perception of the exceptional capabilities of a shop floor system. Probing is discussed in detail through practical examples, leading into in-process gaging. The appendixes are a collection of the most useful codes. Rather than using photographs, this new book uses drawings and sketches in order to provide the interested reader with working information, not easily available, yet applicable to most current systems.

The contents are structured in response to an extensive market survey. The "fundamental" part of the book is written for industrial technology, engineering technology, undergraduate students, junior colleges, and trade schools, as well as for technicians and operators in industry, or for use in courses offered by machine tool manufacturers and distributors. The "advanced" part is a natural followup of the fundamental part. The chapters dealing with user macros, parametric subroutines, computer graphics, and probing, etc., are a unique collection of typical programs. They can be taught as a second-level course, enhanced by applicable questions and problems. The material is introduced incrementally, and the chapters are self-contained with respect to the new material presented. Interesting problems are presented and solved, such as the interfacing of a PC with a tape punch.

All the programs in this book were tested on equipment. The extensive programming content has been checked and rechecked. The reader is, however, encouraged to pretest any program using normal safety procedures.

The authors wish to thank all the readers of *Computer Numerical Control Programming* who took the time and made the effort to contribute numerous suggestions based on their field experience. The various chapters are now concluded with problems and assignments, and an Instructor's Guide includes suggestions and answers to the various assignments.

We would like to acknowledge the many people who have originated the new knowledge, or made it available for inclusion in this book. We would like to express our appreciation to the support afforded advanced technology by the progressive administrators of Humber College. We wish to thank our colleagues, friends, past and present students from the college system, the manufacturing industry, and the machine-tool distributors for their help, cooperation, information contributed or useful suggestions. Some names that come to mind, by no means all, are Ken Palfery, Vice President Manufacturing, and I. "Hank" Ankurs, CNC Manager, McDonnell Douglas of Canada; Andrew Orton and William Kwong from the Technology Division at Humber College; Steve Pereira, Sales Manager, Ferrotechnique Ltd; Ray White, General Manager, Cincinnati Milacron Canada, and Mike Jackson, Cincinnati Milacron, Inc. We are most grateful to the Prentice-Hall editorial group for their continuous encouragement and long-distance support. And finally, we thank Livia Pusztai and Rena Sava, for putting up with more of the same, for so many years.

Michael Sava

Joseph Pusztai

Computer Numerical Control Programming

1

Trends in CNC

The evolution of the machine tool industry could hardly be appreciated without a brief review of its birth and growth. John Wilkinson built his metal-cutting boring machine in the eighteenth century, but nearly two centuries of evolution were needed to produce the hydraulic tracer-controlled copy mills and lathes. The next stage, automation, was brought about by mass production of automobiles, agricultural implements, household appliances, chemical products, as well as inventory and financial data handling. Three kinds of automation met the needs of society for a major part of the twentieth century:

1. Automotive or fixed assembly line automation (Detroit type),
2. Process control automation, primarily used in the manufacture of chemical and food products,
3. Data processing, first developed for processing payrolls, data collection, and inventory control.

The Second World War marked the turning point in the ability of the metal-cutting industry to cope with the requirements facing it. The ambitious aircraft and missile projects of the U.S. Air Force, combined with the demands for commercial jets, made it quite clear that conventional manufacturing could not fulfill future needs. A study of the U.S. government showed that the combined resources of the entire U.S. metal-cutting industry in 1947 could not produce the parts needed by the Air Force alone.

Under contract to the U.S. Air Force, the Parsons Corporation undertook the development of a flexible, dynamic manufacturing system, designed to maximize productivity by emphasizing details required to achieve desired accuracies. This system would allow design changes without costly modifications to tooling and fixturing, and it would fit into a modern, productive manufacturing management for small-to-medium sized production runs. The Parsons Corporation subcontracted the development of the control system to the Massachusetts Institute of Technology (MIT) in 1951. A control, which could be applicable to a wide variety of machine tools, would drive a slide lead screw through an interface, as instructed by the output of a computer. MIT met the challenge successfully and in 1952 demonstrated a Cincinnati Hydrotel milling machine equipped with the new technology, which was named Numerical Control (NC) and used a prepunched tape as the input media. Since 1952, practically every machine tool manufacturer in the Western World has converted part or all of its product to NC.

The first NC machines used vacuum tubes, electrical relays, and complicated machine-control interfaces. The second generation of machines utilized improved miniature electronic tubes and, later, solid-state circuits. As computer technology improved, NC underwent one of the most rapid changes known in history. The third generation used much-improved integrated circuits. Computer hardware became progressively less expensive and more reliable, and NC control builders introduced for the first time Read Only Memory (ROM) technology. ROM was typically used for program storage in special-purpose applications, leading to the appearance of the computer numerical control (CNC) system. CNC was successfully introduced to practically every manufacturing process. Drilling, milling, and turning were performed on "machining centers" and "turning centers." CNC took over glass cutting, pattern making, electrical discharge machining, steel-mill roll grinding, coordinate measuring, electron beam welding, tube bending, drafting, printed circuit manufacturing, coil winding, functional testing, robots, and many other processes.

A set of preprogrammed subroutines, named "canned cycles," were developed for use in routine operations. They were recorded into the ROMs and remained there even after power was shut off. For the first time, this concept made it possible to read the machining program into memory and to operate the machine from memory. In addition to the advantages of editing, the problems caused by erroneous tape reading disappeared.

Along with the many canned cycle options, CNC builders introduced displays for visual editing of part programs in memory. Various in-cycle problems generated alarms and hundreds of diagnostic messages which could be displayed as applicable. Practically every function of the machine was tied into the system and monitored during operation. A constant surface speed control was incorporated into the lathe controllers and continuously anticipated the most efficient spindle speed for the next cut to minimize time lost for spindle acceleration. The conventional linear and circular interpolation in cartesian (rectangular) coordinates were supplemented by polar coordinates and helical interpolation. Safe zones, which could be set through programmed codes or internal parameters, created an electronic crash barrier to prevent tool collision. The latter group of features marked the arrival of high technology to the manufacturing or metal-cutting industry.

The improvement in drives was as important for the system as the contribution of the microprocessor or the minicomputer.

The feed drives, usually known as servodrives, consist of a motor and its control, which receives its motion instructions from the CNC. Their performance is essential to the accuracy, reliability, and flexibility of the CNC system.

The open-loop system is normally used in simple point-to-point, or positioning, systems, although improvements in technology have made it possible to install the system in contouring systems as well. The closed-loop configuration is more accurate and reliable, as reflected by its higher cost.

Although many CNC systems still use hydraulic or pulse motors, the DC drives have gained dominance on a much larger scale. In most cases, the drive packages are purchased from specialized drive system builders. These direct current (DC) permanent-magnet wound field servomotors range from 3,000 revolutions per minute (rpm) to less than 1 rpm without stalling. They develop peak torque capabilities with high slide acceleration and low inertia for optimized system response. Most drive systems offer a choice between transistorized silicone-controlled rectifiers and pulse-width modulation over the full range of amplified voltages. These drives can now drive virtually any lead screw. Their high-response inner current loops provide reliable regulation of torque-load disturbances. They can also be built with high-gain preamplifiers to close high-bandwidth velocity loops. The DC drives provide the answer to the most essential needs of acceleration, deceleration, stopping, and constant velocity, with inherent shaft stiffness for successful operation of the CNC system. The same drive systems actuate robots, transfer lines, flight simulators, graphic plotters, etc. As these drives are infinitely variable and fully regenerative, they can provide for maximum performance and control over the whole range of the motor. By eliminating gearboxes and clutches, the cost of drives for the third-generation CNC systems was reduced substantially.

The fourth-generation microprocessor CNC incorporated in many cases the controversial bubble memory. The bubbles are magnetic garnet crystals grown on nonmagnetic substrate, ranging in size from 2 to 30 micrometres, and used as nonvolatile data storage. Although at this stage it is not competitive in the large computers, the bubble memory is closing the cost gap with disk storage devices. Insensitive to adverse temperature changes, dust, and vibration, the bubble memory has demonstrated superior reliability in shop environment. General Numerics introduced its fourth-generation CNC using bubble memory. More rugged than the others, bubble memory is still expensive compared to hard disk, and slow compared to ROM.

Among the strengths of the fourth-generation microprocessor CNC (MCNC) are added part program memory storage, reduction of printed circuit boards, programmable interface, faster memory access, parametric subroutines, and macro capabilities.

The system user can now write specific canned cycles directed to particular applications ("user macros"), far more economical and efficient than conventional canned cycles. Mathematical calculations with do-loop subroutines using variables can now be incorporated in the part program. The microprocessor controls both computations and motion commands. Thus, following an in-process gaging, an out-of-tolerance condition will be fed back, and the tool offset will be automatically modified to achieve the desired part dimensions.

The fourth-generation microcomputer CNC system has the ability to control typical robot functions such as loading and unloading parts. Using the teach-in learning mode, the robot can be programmed to change tools or to remove chips.

Where will technology go from here? To a large extent it depends on the knowledge of the system users and the demands they will pose to the designers and builders of manufacturing systems. CNC will probably remain for a long time one of the most practical elements of computer-aided design and computer-aided manufacturing technology.

2

Mathematics
for Computer
Numerical Control

The previous chapter emphasized the extraordinary capabilities of the CNC machine. However smart a CNC may be, it simply cannot think. It can perform unlimited numbers of activities and can repeat any number of operations continuously and consistently by the use of numeric directions and commands. In this chapter we will discuss two different types of mathematics: one which deals with and leads up to the tape codes, or Binary Coded Decimal (BCD) mathematics; and the other which is used by the programmer as an aid to calculate tool centerline dimensions. The latter type of mathematics, known as trigonometry and analytical geometry, will present useful formulas essential to every CNC programmer for writing efficient part programs.

2.1 THE NUMERICS THAT CNC MACHINE TOOLS UNDERSTAND

The basic hardware of the CNC consists of the input units, the computing or mathematics unit, the memory unit, the control unit, and the output units. The function of any input unit is to provide data to the computer in the form of numeric instructions. Present CNC systems are designed to operate with different input media. The most common of these is the punched tape, mainly because it can be read inexpensively, is less sensitive to handling, is inexpensive to purchase, and requires less equipment to make and less costly space for data storage. Its disadvantage, however, is that it cannot be reused. The magnetic tape has limited use as a CNC media, it requires sophisticated (expensive) equipment for program recording and reading, and the programmer or operator cannot see the recorded codes and therefore cannot read them. Recording errors are not as obvious and visible as they are in punched tape. Magnetic tape requires special storage space and must be handled carefully to avoid erasing the program. The typewriter, more commonly known as the keyboard, has limited use because of the operator's speed. It may not be used for long program input, but is used primarily for small programs. Its main use is to edit (correct) programs already in memory or to generate single operations in the Manual Data Input (MDI) mode. The objective here is to introduce the reader to the evolution of the codes we use in our input units to communicate with the CNC computer.

The part program, once read into the computer memory, becomes a set of instructions to carry out specific commands which may be preparatory in nature or auxiliary to the machining process, in-process gaging, or just simple calculations to support the machining process. The most popular CNC memories are still the Semiconductor IC memory and the Magnetic Core memory. However, "bubble" memories are being used now in some CNC systems. The internal CNC memory can only handle small amounts of data, but at a very fast rate. Because of this rate of speed, CNC can perform linear, circular, and parabolic interpolation (calculations) at a rate of 200 to 300 inches per minute (ipm) slide velocities.

The "brain" or control unit of the CNC controls how these operations are performed. It translates the memory instructions and specifies what operations are to be performed in what sequence. The mathematical unit performs simple addition, subtraction, multiplication, and division functions. The results are fed back to the memory for storage or read out to the various output units of the CNC. These output units are servodrives for slides or program readouts to teletype printer, CRT, or tape punch unit. Tool changes or other miscellaneous codes are not handled in the mathematics. The codes or numbering system of these calculations are different from the input code used in the punched tape. Later in this chapter we will discuss in detail how the different numbering systems function in the CNC.

The CNC is a special-purpose computer, using special CNC commands in a simplified manner called programming. These commands are written instructions in schematic form. The programmer does not have to describe in every detail what steps the CNC is to perform. A few commands such as we use to instruct the machine to cut a 360° circular arc will cause the computer to perform thousands of calculations involving additions, subtractions, multiplications, and divisions. A CNC that cannot do these types of calculations would have very little use in complex parts manufacturing.

The accuracy of the calculations is limited by the digits used for each number. In most CNCs the number length, called the number of binary digits, is fixed to 8 and 16 bits (binary digits). The precision of the CNC far exceeds the physical limitations of the mechanical devices such as lead screws and slides. In spite of all the glorious things said about the CNC, it is the programmer who does the thinking and achieves the precision. The CNC is a very primitive piece of hardware; it can only understand numbers composed of "1" and "0," in electrical terms "on" or "off," sensing the presence or absence of magnetism or voltage.

2.1.1 Numbering Systems: Decimal Number System

In our everyday life we seldom think of, or analyze, the numbers we use and work with. For example, the number 649 should really be written as $649_{(10)}$ meaning base 10. The base 10 system has digits from 0, 1, 2, 3, 4, . . . 9, but there is no 10. The 10 is not a basic digit in the system. The second important point to note is that the position of the digit in the number defines the value of units, tens, hundreds, thousands, etc.

The number

$$649_{(10)} = 9 \times 10^0 + 4 \times 10^1 + 6 \times 10^2$$

$$= 9 \times 1 + 4 \times 10 + 6 \times 100$$

$$= 649_{(10)}$$

In general terms any number (N) can be expressed by the following general equation:

$$N = d_n R^n + d_{n-1} R^{n-1} + \cdots \cdots + d_2 R^2 + d_1 R^1 + d_0 R^0$$

where N is the number, d_n is the digit of nth position, and R^n is the base or radix of the nth position.

Since computers are simple electronic devices that can only sense voltage on (1) or off (0), a light being on (1) or off (0), a transistor on (1) or off (0), or magnetic field on (1) or off (0), they cannot work with the decimal system's complexity.

2.1.1.1 Binary number system.   A numbering system that is made up of only the two basic digits "0" and "1" is called base 2 or binary number system. This is the basic system that computers work with; it is also the basis for our punched tape codes.

Comparing the decimal and binary bases with their powers, we find no difference in the principle:

$$10^0 = 1 \qquad\qquad 2^0 = 1$$

$$10^1 = 10 \qquad\qquad 2^1 = 2$$

$$10^2 = 100 \qquad\qquad 2^2 = 4$$

$$10^3 = 1000 \qquad\qquad 2^3 = 8$$

The latter is the base our punched tape codes of numerics operate on. The binary numbers can now be handled by on-off type of electronic circuits. The equivalence between decimal and binary numbers is shown in Table 2.1.

TABLE 2.1   DECIMAL AND BINARY
NUMBERS

Decimal	Binary	Decimal	Binary
0	0000	11	1011
1	0001	12	1100
2	0010	13	1101
3	0011	14	1110
4	0100	15	1111
5	0101	16	10000
6	0110	17	10001
7	0111	18	10010
8	1000	19	10011
9	1001	20	10100
10	1010		

2.1.1.2 Converting decimal to binary

Example

Convert Decimal 327 to Binary.

Solution	**Remainder**	
2|327	1	(LSD) least significant digit
2|163*	1	
2|81*	1	
2|40*	0	
2|20*	0	
2|10*	0	
2|5*	1	
2|2*	0	
|1*	1	(MSD) most significant digit

Read from the MSD to the LSD, the binary equivalent of 327 is 101000111. (NOTE:
* = *quotients.*)

Example

Convert Decimal 92 to Binary.

Solution	**Remainder**	
2|92	0	LSD
2|46	0	
2|23	1	
2|11	1	
2|5	1	
2|2	0	
1	1	MSD

Read from the MSD to the LSD, the binary equivalent of 92 is 1011100.

2.1.1.3 Converting binary to decimal.   Changing the base from 10 to 2 in the general equation discussed under the decimal section is a very simple operation.

Example

Determine the decimal value of the following binary numbers:

1. $(1011)_2 = (?)_{10}$

Solution

- Assign powers $0, 1, 2, 3$, to the binary numbers from right to left (these powers are for the base).

$$1^3 \quad 0^2 \quad 1^1 \quad 1^0$$

- Substitute these binary numbers into the general equation using the base or radix 2 instead of 10, at the power corresponding to the location of the digit, as shown above, and multiply each one by the corresponding binary digit.

$$N = 1 \times 2^3 + 0 \times 2^2 + 1 \times 2^1 + 1 \times 2^0$$
$$= 8 + 0 + 2 + 1 = (11)_{10}$$

2. $(101000111)_2 = (?)_{10}$

Solution

- $1^8 0^7 1^6 0^5 0^4 0^3 1^2 1^1 1^0$

$$N = 1 \times 2^8 + 0 \times 2^7 + 1 \times 2^6 + 0 \times 2^5 + 0 \times 2^4$$
$$+ 0 \times 2^3 + 1 \times 2^2 + 1 \times 2^1 + 1 \times 2^0$$

- Canceling terms with zero digits

$$N = 1 \times 2^8 + 1 \times 2^6 + 1 \times 2^2 + 1 \times 2^1 + 1 \times 2^0$$
$$= 256 + 64 + 4 + 2 + 1 = (327)_{10}$$

2.1.1.4 Fractional binary numbers.   Since most programmed numbers are fractions of a whole (decimal), fractional numbers are as important as integers. The method of converting fractions to binary numbers differs from the integer method. Instead of dividing, we multiply the fraction by 2. The number to the left of the decimal point of the product will be the binary number, while the sum to the right of the decimal point will be used as multiplicand. This multiplication is repeated until the desired accuracy is attained.

Example

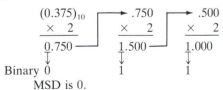

```
(0.375)₁₀    ──────▶ .750      ──────▶ .500
 ×    2              ×    2              ×    2
 0.750 ─────┘        1.500 ─────┘        1.000
    │                   │                   │
    ▼                   ▼                   ▼
Binary  0               1                   1
     MSD is 0.
     The answer is (0.011)₂.
```

Converting fractional binary to decimal is identical to the integer conversion. The general formula we have used for integers used positive powers of the base. For fractions, these powers will have to be changed to negative as:

$$N = d_1 \times R^{-1} + d_2 \times R^{-2} + d_3 \times R^{-3} + \cdots \cdots + d_n \times R^{-n}$$

Example

$$(0.011)_2 = (\ ?\ )_{10}$$

Solution

• Assign powers,

$$0^{-1}1^{-2}1^{-3}$$

These powers, as in our previous example, are transferred to the base as:

$$N = 0 \times 2^{-1} + 1 \times 2^{-2} + 1 \times 2^{-3}$$

Canceling the zero terms and performing the summation as shown below:

$$1 \times 2^{-2} = 1 \times \frac{1}{2^2} = \frac{1}{4} = 0.25$$

$$1 \times 2^{-3} = 1 \times \frac{1}{2^3} = \frac{1}{8} = 0.125$$

$$\therefore \quad 0.25 + 0.125 = (0.375)_{10}$$

Mixed numbers can convert just as easily from binary to decimal if we remember that the powers of the base are positive to the left and negative to the right of the decimal point.

Example

Convert $(1011.1011)_2$ to decimal.

Solution:

$$1^3 0^2 1^1 1^0.1^{-1} 0^{-2} 1^{-3} 1^{-4}$$

$$N = 1 \times 2^3 + 0 \times 2^2 + 1 \times 2^1 + 1 \times 2^0 + 1 \times 2^{-1} + 0 \times 2^{-2}$$

$$+ 1 \times 2^{-3} + 1 \times 2^{-4}$$

$$= 1 \times 2^3 + 1 \times 2^1 + 1 \times 2^0 + 1 \times 2^{-1} + 1 \times 2^{-3} + 1 \times 2^{-4}$$

$$= 8 + 2 + 1 + 0.5 + 0.125 + 0.0625$$

$$= (11.6875)_{10}$$

Example

Convert $(29.1875)_{10}$ into binary.

Solution:

- Convert integers

$2\lfloor 29$ 1 ← LSD
$2\lfloor 14$ 0
$2\lfloor 7$ 1
$2\lfloor 3$ 1
$2\lfloor 1$ 1   MSD   Integer answer is 11101

- Convert the decimal digits as:

$$
\begin{array}{cccc}
0.1875 & .3750 & .7500 & .5000 \\
\times\ \ \ 2 & \times\ \ \ 2 & \times\ \ \ 2 & \times\ \ \ 2 \\
\hline
0.3750 & 0.7500 & 1.5000 & \\
\downarrow & \downarrow & \downarrow & 1.0000 \\
0 & 0 & 1 &
\end{array}
$$

Decimal answer is 0.0011

$\therefore (29.1875)_{10} = (11101.0011)_2$

There are other number systems used in computer technology, such as the base 8 $(N)_8$ called octal, and the base 16 $(N)_{16}$ or hexadecimal, in addition to the binary number system. Detailed discussion of other number systems would be beyond the scope of our objectives.

However, a brief discussion of the four basic arithmetic operations with binary numbers will be a useful aid for the reader (programmer). In each operation, the reader must "memorize" only four combinations, as opposed to the decimal system where one had to "memorize" 100 combinations. For addition in binary numbers, see Table 2.2; Table 2.3 shows subtraction in binary numbers.

2.1.1.5 Addition

Example

```
                1  1  1  1  1      ←——o carries    Note:    1
    A      +       1  1  1  1           15               + 1
    B             1  1  1  1  1        + 31             0, carry 1
    ─────────────────────────────
    A + B    1  0  1  1  1  0        (46)₁₀
```

TABLE 2.2 ADDITION IN BINARY NUMBERS

Augend		Addend		Sum	Carry
0	+	0	=	0	0
1	+	0	=	1	0
0	+	1	=	1	0
1	+	1	=	0	1

TABLE 2.3   SUBTRACTION IN BINARY NUMBERS

Minuend		Subtrahend		Difference	Borrow
0	−	1	=	1	1
1	−	1	=	0	0
1	−	0	=	1	0
0	−	0	=	0	0

2.1.1.6 Subtraction

Example

```
Borrows      000000
   A         11001      49
   B       − 10000    − 32
A − B        010001    (17)₁₀
```

Subtraction of large binary numbers is difficult, especially for those who are not familiar with the binary number system. Some of you may find it easier to work with another method called Binary Complement Subtraction. The method is described below, step by step, using the same example as above.

1. A   49   1 1 0 0 0 1   1 1 0 0 0 1
   B − 32 − 1 0 0 0 0 0   + 0 1 1 1 1 1
        17

2. Write down the minuend as shown on the right. Then write down the complement of the subtrahend below.

3. Instead of subtracting, add the two numbers. Once the addition is completed, the carry must be added to the sum as shown:

```
1  1  1  1  1  1      carry
  1  1  0  0  0  1           49
  0  1  1  1  1  1         − 32      Complement of 32
  0  1  0  0  0  0
(1)────────────→1
     1  0  0  0  1         (17)₁₀
```

(NOTE: *The complement of binary 0 is 1 and the complement of binary 1 is 0.*)

2.1.1.7 Binary multiplication.

Most CNC systems do not perform multiplications. The multiplication is implemented by repeated addition, the same way as the addition of all partial products is performed to obtain the final sum (see Table 2.4). The formation of partial products is easy (the same as the decimal multiplication). The addition of all the partial product is more difficult. You must count the number of ones in the column: If it is even, the sum of the column is 0. If it is odd, the sum of the column is 1. For every pair of 1s there is one carry to the next higher position.

TABLE 2.4 MULTIPLICATION IN BINARY NUMBERS

Multiplicand		Multiplier		Product
1	*	1	=	1
0	*	1	=	0
1	*	0	=	0
0	*	0	=	0

Example

$$\begin{array}{r} 43 \\ \times\ 22 \\ \hline 946 \end{array} \quad \begin{array}{l} \text{A} \\ \text{B} \end{array} \quad \begin{array}{r} 1\ 0\ 1\ 0\ 1\ 1 \\ \times 1\ 0\ 1\ 1\ 0 \\ \hline \end{array}$$

Solution

```
        0  1  1  1, 1  1  1, 0  0    carry
                 )  *         )  *        ↓  multiply by "0"
                    0  0  0 ) 0  0  0   start with LSD
                 1/ 0  1, 0 / 1) *1    —multiply by 1
              1, 0  1, 0 ) *1  1/       —multiply by 1
        0  0) *0  0) *0) 0               —multiply by 0
        1  0  1  0  1  1                  —multiply by 1
A · B = 1  1  1  0  1  1  0  0  1  0   add
```

(NOTE: $\frac{1}{1}$)* = 0 *with carry of* 1.)

The only possible way to verify a large binary number such as our result is through binary-to-decimal conversion (discussed in the preceding pages).

$$N = 0 \times 2^0 + 1 \times 2^1 + 0 \times 2^2 + 0 \times 2^3 + 1 \times 2^4 + 1 \times 2^5 + 0 \times 2^6$$

$$+ 1 \times 2^7 + 1 \times 2^8 + 1 \times 2^9$$

$$= 0 + 2 + 0 + 0 + 16 + 32 + 0 + 128 + 256 + 512$$

$$= (946)_{10}$$

2.1.1.8 Binary division.   As in the case of multiplication, most CNC systems perform divisions by repeated subtraction of the divisor from the dividend. The division rules for two 1-bit binary numbers are shown in Table 2.5.

Example

Dividend Divisor Quotient

$$26 \div 5 = 5.2$$
$$10$$
$$0$$

Solution $11010 \div 101 = 101.001$

```
11010 ÷ 101 = 101.001
101
‾‾‾‾
00110
 101
 ‾‾‾‾
 001000
   101
   ‾‾‾‾
    11
```

TABLE 2.5 DIVISION IN BINARY NUMBERS

Dividend		Divisor	Quotient	Remainder
1	÷	1	1	0
1	÷	0	Undefined	Undefined
0	÷	1	0	1
0	÷	0	Undefined	Undefined

EXERCISES

2.1.1. Convert the following decimal numbers to binary:

(1) 23 D		**(5)** 379 D		**(9)** 22 D
(2) 69 D		**(6)** 127 D		**(10)** 88 D
(3) 147 D		**(7)** 79 D		**(11)** 222 D
(4) 1897 D		**(8)** 149 D		**(12)** 3638 D

2.1.2. Convert the following binary numbers to decimal:

(1) 10101011 B	**(4)** 11111101 B	**(7)** 111111 B
(2) 1001111 B	**(5)** 11101 B	**(8)** 100001 B
(3) 1011 B	**(6)** 00101 B	**(9)** 110100 B

2.1.3. Convert the following fractional decimal numbers to binary:

(1) 0.6565 D	**(3)** 0.737 D
(2) 0.8759 D	**(4)** 0.4375 D

2.1.4. Convert the following fractional binary numbers to decimal:

(1) 0.111011 B	**(2)** 0.110101 B

2.1.5. Convert the following decimal numbers to binary:

(1) 6.875 D	**(3)** 3.225 D	**(5)** 13.4315 D
(2) 11.235 D	**(4)** 6.235 D	**(6)** 22.555 D

2.1.6. Add the following binary numbers:

(1) 11001 + 111	**(2)** 00110 + 1010	**(3)** 111111 + 111

2.1.7. Subtract the following binary numbers:

(1) 110010 − 110000	**(3)** 110011 − 01010	**(5)** 10000 − 00110
(2) 101111 − 1011	**(4)** 11101 − 0101	**(6)** 11011 − 10100

2.1.8. Multiply the following binary numbers:

(1) 11111 × 1110	**(3)** 111 × 101	**(5)** 01110 × 011
(2) 101101 × 1011011	**(4)** 1111 × 010	**(6)** 1100 × 1010

2.1.9. Divide the following binary numbers:

(1) 11011 ÷ 1011	**(3)** 11011 ÷ 110	**(5)** 001101 ÷ 011
(2) 1011 ÷ 111	**(4)** 1111 ÷ 110	**(6)** 10101 ÷ 111

2.1.2 Binary-Coded-Decimal Code (BCD)

We have earlier established that the CNC system is an electronic device that can understand simple "on" (1) or "off" (0) states. We have also showed that the base

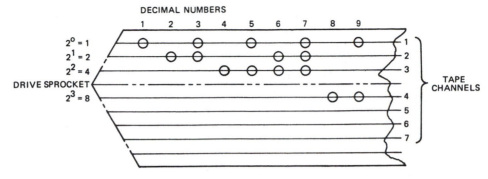

Figure 2.1   BCD tape code.

2 or binary number system allows us to represent any decimal number in binary. All CNC systems use some sort of binary system for their arithmetic or internal operation, but externally the real world works with the decimal system. We have seen that conversion between decimal and binary can be long and erroneous for large numbers. As a compromise, a binary-coded-decimal (BCD) coding was developed, based on the position of the numbers used to describe our CNC tape codes. The first, second, third, and fourth positions can be described as:

$$2^0 = 1, 2^1 = 2, 2^2 = 4 \text{ and } 2^3 = 8$$

Reading from right to left, the weight can be written as 8-4-2-1. For this reason, this system is often called the 8421 code. This code compresses the binary numbers so that they can be punched in tape to control our CNC system. The BCD (8421) codes are punched in rows across the 1-inch-wide standard tape; each row represents one digit in the tape and successive rows can represent any numbers. Figure 2.1 illustrates the BCD tape code principle.

One of the main advantages of the BCD system is that once learned it is easy to read the values represented by punched holes. For example:

Digit 1 is represented by a hole in channel 1 ($2^0 = 1$)
Digit 2 is represented by a hole in channel 2 ($2^1 = 2$)
Digit 5 has no code of its own, but is the sum of $2^2 = 4$ plus
$2^0 = 1$ hence $4 + 1 = 5$.

The reader can easily visualize the simplicity. The BCD codes are used in both Electronic Industries Association (EIA) and the American Standard Code for Information Interchange (ASCII) Systems, which will be discussed in more detail in the following pages.

2.2 CNC TAPES

When punched or recorded (in the case of magnetic tape), these tapes are predominantly used by CNC systems as input/output or control media. The reader may find many different makes and colors of 1-inch-tape commercially available. All tapes are manufactured to an EIA standard (shown in Figure 2.2), which also outlines the tolerances required by the manufacturers of tape punching and reading equipment. Selection of the tape material should be based on the type of tape reader and tape punching unit available.

For mechanical tape readers (few in new systems), mylar base tapes should be used because mylar provides considerably longer life under repeated rereading than the regular paper tape. Although it is more expensive, it does not require special punching facilities, has low wear rate, and provides excellent resistance to oil and grease.

For photoelectric tape readers and systems with memory, the inexpensive paper tape should suffice. The primary requirement is to provide high opacity and low reflectivity. However, tape readers with 500 characters per second (cps) or higher reading speeds may damage this tape if repeated readings are required. Because of the high acceleration and deceleration rates, these tape readers are most reliable with laminated paper-mylar or a less costly aluminized-mylar tape.

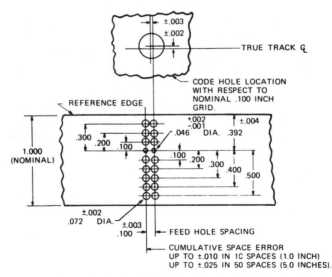

Figure 2.2 Electronic Industries Association Standards for 1-inch perforated tape. (Courtesy Electronic Industries Association)

	MAG TAPE CODING							(ISO) ASCII								EIA							
CHARACTER	1	2	3	4	5	6	7	1	2	3	• 4	5	6	7	8	1	2	3	• 4	5	6	7	8
0																							
1																							
2																							
3																							
4																							
5																							
6																							
7																							
8																							
9																							
a																							
b																							
c																							
d																							
e																							
f																							
g																							
h																							
i																							
j																							
k																							
l																							
m																							
n																							
o																							
p																							
q																							
r																							
s																							
t																							
u																							
v																							
w																							
x																							
y																							
z																							
+																							
−																							
/																							
.																							
,																							
%																							
&																							
(																NOT		ASSIGNED					
)																NOT		ASSIGNED					
:																NOT		ASSIGNED					
CR																							
DELETE																							
SPACE																							
=																							
TAB																							

Figure 2.3 ASCII (ISO) and EIA punch tape codes used by CNC controls.

2.3 TAPE-PUNCHING FACILITIES

Several different manual tape punches are commercially available. However, with the drastic price reductions in the minicomputer field, we recommend the purchase of a minicomputer based tape-punching facility that works with a floppy disk drive.

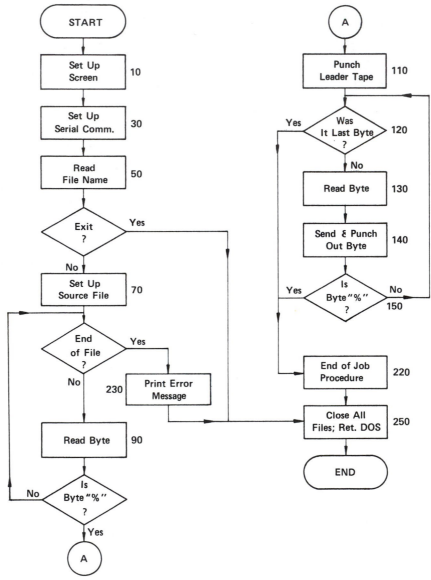

Figure 2.4   Interfacing a PC to a tape punch.

The writers are using an IBM PC coupled with a tape punch and printer. This system provides an inexpensive trouble-free high-speed tape punch; the program is written on the screen and can be edited (corrected) before a tape is prepared. The diskette provides the lowest cost storage device available as dozens of programs can be recorded on a single diskette.

For computer-aided programming, we should mention that editing software can be purchased with the hardware for most commercially available mini-computers.

CNC systems use two different punched-hole codings: the EIA RS-244 standard was developed and used predominantly by the North American industry; ASCII RS-358 standard was developed in the United States but is accepted and used throughout the world under the name ISO. While some older NC controls work with EIA, most current CNC systems accept both punched codes.

Parity Code.   To minimize the possibilities of errors during the internal handling of binary data within the CNC control, as well as during tape punching and reading, a Parity Check system has been implemented in the standardized coding.

The Parity Check Code for the EIA consists of an extra hole in the fifth row, in addition to the BCD code if the holes would otherwise be even, so that the number of holes across the tape will always be "odd." Holes in track five are used exclusively for this purpose. The Parity Check Code for the ASCII requires "even" numbers of holes across the tape in every row. Holes in track eight are used exclusively for parity adjustment whenever it is required. Punched tape codes were also developed for alphabetic and other symbolic keyboard codes by ASCII. Both ASCII and EIA punch tape codes are illustrated in Figure 2.3.

2.3.1 Interfacing a PC to a Tape Punch

The following program will cause a program written and edited on the screen to be punched by a tape punch, serially connected to the PC.

PROGRAM

This program can be written in a number of languages. BASIC was selected due to its widespread acceptance.

10 SCREEN 0.0 : WIDTH 80	Text mode, black and white, 80 characters/line
20 KEY OFF : CLS : CLOSE : LOCATE 5	Opening statement. Turn off function keys, clear screen, assure that all files are closed, position cursor.

30 OPEN "com2 : 300, e, 7, 1 cs, ds, cd" AS #1	Communication set up statement. Baud rate 300 bits per second, parity even, 7 data bits, 1 stop bit, RS 232 format serial transmission. The statement is a string labeled #1. This is the file the program will write to, or the output file.
40 LOCATE 18,5	Position cursor at the intersection of row 18 and column 5.
50 INPUT " Enter file name or 'EX' to exit =>";F$	The answer to the prompt is stored as string data F$.
60 IF F$ = "EX" OR F$ = "ex" THEN 250	The programmer wishes to exit program. Control is transferred to sequence 250 and process is terminated.
70 OPEN F$ FOR INPUT AS #2	A new file, labeled #2, was opened. This file, containing the answer to the initial prompt, is the one the program will read from, or the input file. Label F$ was assigned in sequence 50.
80 IF EOF(2) THEN 230	This line checks the input file (#2) for an End of File condition, to avoid an "input past end" error. If end of file was reached without reading a "%," an error has taken place and an error message is required. Control is transferred to seq. #230.
90 X$ = INPUT$(1,#2)	The program will read 1 byte from the input file (#2). The byte read will be stored as string data X$.
100 IF X$<> "%" THEN 80	The tape file must start with a %. This causes the program to ignore anything prior to %, such as man-readable messages. If the byte read is not %, control is transferred to seq. 80, end of file rechecked, and a byte read.
110 GOSUB 170	The byte read was %. The program is allowed to proceed in sequence, and control is transferred to subroutine, at seq. 170. Leader tape is punched.
120 IF EOF(2) THEN 220	End of input file is checked again. If it was reached, control is transferred to 220, where the process of punching trailer tape will be initiated, then the operation will be terminated.
130 X$ = INPUT$(1,#2)	End of file was not reached. The program is allowed to proceed. The next byte is read from input file #2.
140 PRINT #1,X$;	The byte read was stored as X$ in the previous line. It is now sent to the output file #1, i.e., the communication line (the interface) as defined in seq. 30, and punched to tape.

150 IF X$ = "%" THEN 220	If the byte read was %, the program is now legitimately ended, and control is transferred to 220 as in seq. 120.
160 GOTO 120	This is the end of the input file reading loop, operating in the range 120 to 160. The loop will check for end of file (CNC program), read the next byte, punch it to tape, check for %, etc.
170 FOR A = 1 TO 200 180 PRINT #1,CHR$(0) 190 NEXT A 200 RETURN	The sequence 170 to 200 is the leader, respectively trailer punching subroutine. 200 characters of "zero ASCII" are sent to the punch (output file #1). Accordingly, 200 lines of blank tape (20 inches) are put through the printer before and after the CNC program being punched.
220 GOSUB 170 : PRINT " End of Punching " : CLOSE : SYSTEM	Closing statement. This instruction transfers command to the subroutine and the trailer tape is punched. The programmer is advised on screen that the process is over. All files opened by this program (#1, #2) are closed, and control is returned to DOS.
230 CLS : LOCATE 10 240 PRINT " Check for missing % sign in your program".	Error message : Sequences 230 and 240 advise the programmer on the screen that "%" is missing from their program.
250 CLOSE : SYSTEM	Close all files and return control to DOS. This line is used in conjunction with the programmer's "Exit" request and the error message advisory.

EXERCISES

2.3.1. Identify the code and decode the following punched tape codes:
   (1)

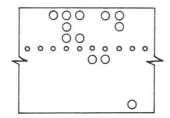

Figure 2.5 Punched tape.

(2)

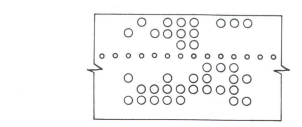

Figure 2.6 Punched tape.

2.3.2. Using the ASCII code, draw the hole pattern of the following tape blocks:
  (1) N11G01M01X1.3551Y-2.005CR
  (2) N09G17S1137X2.500Y6.975M03F09CR

2.4 MAGNETIC TAPE CODES

Magnetic tape codes are also derived from the BCD system and in most CNC applications they are recorded (instead of punched) on half-inch-wide cartridged magnetic (mag) tape. The codes or data are recorded in seven parallel channels or tracks (Figure 2.3), using a character density of 800 to 4,800 per inch. Their reading speeds are expressed in inches per second (ips) or characters per second (cps). This speed is normally 160 ips or 45,000 cps on most CNC controls. The read-write heads may be of one- or two-gap types. The one-gap head is used for either reading or writing, but only one at a time (more popular with CNC); the two-gap type can write a code (bit) and read it back, while the bit is still under the head, for parity check.

Magnetic tape media was used by several U.S. control manufacturers during the late 1960s and 1970s, mostly by Bendix, Thompson-Rand-Wooldridge, and Kearney & Trecker for four- and five-axis contouring. In spite of the high reading characteristics and the lower basic cost of the control, mag tape never gained the wide acceptance that punched tape did. The codes cannot be seen by the naked eye; however, special optical viewing instruments are available on the market to view the recorded codes, as shown by Figure 2.3. The alphanumeric codes are shown by dashes (-). Interactive CNC controls with mag tape storage are periodically coming to the market.

2.5 MATHEMATICS FOR THE PROGRAMMER

The question most asked by persons wishing to learn CNC programming is "What level of mathematics do I need to be able to learn CNC programming?" Unfortunately there is no simple answer. However, it is safe to state that a part programmer for lathe and/or two-axis contour milling should have a working knowledge of coordinate systems, trigonometry, analytical geometry, and cutting forces. While the objective of this book is not to discuss lengthy mathematical derivations, it will provide valuable information for those who wish to review areas of concern.

2.5.1 Cartesian Coordinate System

Most of you are familiar with the rectangular or cartesian coordinate system you have learned in high school. All the CNC systems are built to function and therefore must be programmed in terms of a coordinate system. The mathematics discussed here will be shown in coordinate systems whenever possible. A two-axis coordinate system is formed by two intersecting straight lines perpendicular to each other (see Figure 2.7), hereafter called X and Y axes.

The sample programs in the book will refer to this two-axis system for positioning and contouring. Drawing a third line, as shown by Figure 2.7, perpendicular to the plane formed by the X-Y axis through the intersecting point will form a three-axis coordinate system. The intersection point is called the "Origin," and the third or "Z" axis will be called the "tool" axis.

Point.   The simplest element is a point (PT) and it can be defined by its X-Y coordinates as PT1 (X, Y), shown by Fig. 2-7 using actual values as PT1 (8.5, 11). This notation refers to a two-axis coordinate system. PT2 and PT3 cannot be defined by this method because PT2 has no X and PT3 has no Y values. These points must therefore use a three-axis notation in the form of PT (X, Y, Z). This notation allows us to define the location of any point in space in terms of our coordinate system as PT2 (0, 6.2, 7), PT3 (5, 0, 6), and PT4 (8.5, 11, 2).

Line.   Any line can be defined by two points in a cartesian coordinate system. (There is another definition using a radius from the origin and an angle measured from the positive X-axis which will be discussed under Polar Coordinates.)

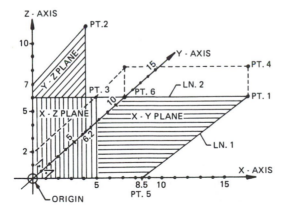

Figure 2.7 Cartesian coordinate system.

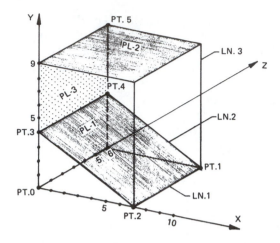

Figure 2.8 Elements of geometry in a cartesian coordinate system.

Example

Define Line 1 (LN1) and Line 2 (LN2).

Solution

PT1 (8.5, 11, 0) is defined in Figure 2.7.
Define the second point, PT5 as PT5 (8.5, 0, 0)
      Now Line 1 can be defined as LN1 (PT1, PT5), or a line going through points PT1 and PT5.

Similarly, LN2 (PT1, PT6), where PT6 (0, 11, 0).

Plane.   Better known in industry as surface, it can be defined by three points. It can also be defined several other ways. However, plane definitions by rotation and transfer are beyond the objective of this book. Some planes, such as PL1, PL2 or PL3 are illustrated in Figure 2.8.

Example

Define Plane 3.

Solution   Define the three points required for the plane definition.

PT3 (0, 4, 0), PT4 (0, 4, 6), PT5 (0, 9, 6)

Define the plane as PL3 (PT3, PT5, PT4).

2.5.2 Polar Coordinate System

The nomenclature of the axis (X-Y-Z) is identical to the cartesian system. However, the coordinate location of the point, line, or plane is defined in terms of a radius (distance from origin to a point) and the angle between the positive X-axis and

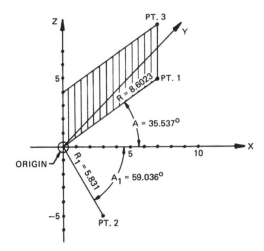

Figure 2.9 Polar and/or cylindrical coordinate system.

the geometric shape we wish to specify. The angle is positive ($+$) in the counterclockwise direction (CCW) and negative when measured in the clockwise (CW) direction from the X-axis. See Figure 2.9.

Some of the latest CNC controls are programmed in terms of polar coordinates. This simplifies the calculations when holes have to be drilled on a circular pattern.

Example

Define the location of PT1 in terms of polar coordinates.

Solution

PT1 (R, A)
PT1 (8.6023, 35.537)

Points not located in the reference plane are defined by their "cylindrical coordinates." PT3 must therefore be defined in terms of its radius "R," angle "A," and height "Z," in the form of PT3 (R, A, Z). Using the dimensions from Figure 2.9, the answer will be PT3 (8.6023, 35.537, 4.0).

A typical CNC application of the cylindrical coordinate system is illustrated in Figure 2.10.

In order to mill the cam groove on the cylinder (centerline of groove shown on drawing), we need first to define the start (PT1) and end (PT2) points in terms of the radius, angle, and height dimensions. The tool path from $-120°$ to $+110°$ describes the rotation of plane 1 to plane 2 position. The rotation plane X-Y is circular, and the third axis motion (Z) is linear. In fact, the tool point will describe a helical motion along the surface of a perpendicular cylinder.

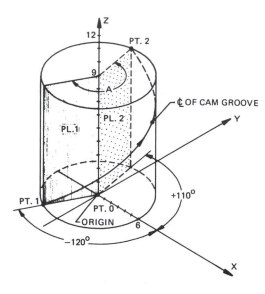

Figure 2.10 Cylindrical coordinate system.

EXERCISES

2.5.1. Using Figure 2.8, define the location of the following geometry:
  (1) PT1   **(3)** LN2
  (2) LN1   **(4)** PL1

2.5.2. Draw the following lines in a coordinate system:
  L1 = P1 (0,4,0), P2 (4,4,0)
  L2 = P3 (4,0,0), P4 (0,0,0)
  L3 = P5 (4,0,2), P6 (0,0,2)
  L4 = P7 (0,4,2), P8 (4,4,2)

2.5.3. Using the geometry of the previous problem, draw the following planes:
  (1) PL1 (L1, L3)   **(2)** PL2 (L2, L4)

2.5.4. Draw a plane using the following points:
  P1 (0,0,0), P2 (0,4,2) and P3 (4,4,0)

2.5.5. Using polar coordinates, define the location of PT2 in Figure 2.9.

2.5.6. Using the cylindrical coordinate system in Figure 2.10, define the location of the following points:
  (1) PT1   **(2)** PT2

2.5.7. Draw the following lines in a polar or cylindrical coordinate system:
  (1) L1 = P1 (6.5, 37.5), P2 (0.75, 185)
  (2) L2 = P3 (16.5, 180), P4 (3.5, 90)
  (3) L3 = P5 (6.0, 88), P6 (0, 45)

2.6 TRIGONOMETRIC FUNCTIONS

The science of "triangle measurement" is commonly known as "trigonometry." Trigonometric functions such as angles, sides of right-angle triangles, and their relationships will be discussed in this section.

2.6.1 Pythagorean Theorem

The square of the hypotenuse in a right-angle triangle is equal to the sum of the squares of the other two sides. See Figure 2.11.

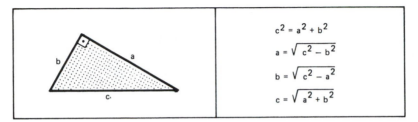

Figure 2.11   Pythagorean Theorem.

2.6.2 Similar Triangles

If the sides of any angle are intersected by two parallel straight lines, two similar triangles are formed. The ratios of the sides can be expressed as shown by Figure 2.12. This relation will hold for any number of parallel lines traced, i.e., a_2, b_2, c_2-a_3, b_3, c_3, etc.

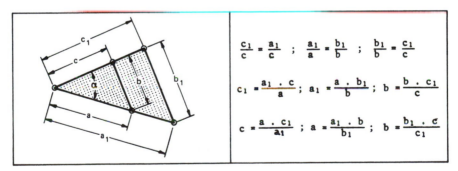

Figure 2.12   Similar triangles.

2.6.3 Sine and Cosine Functions

We will only show the most frequently used functions (Figure 2.13).

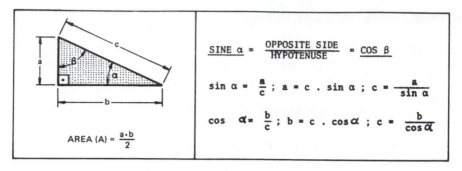

Figure 2.13   Sine and cosine functions.

2.6.4 Tangent and Cotangent Functions

The most frequently used functions are illustrated in Figure 2.14.

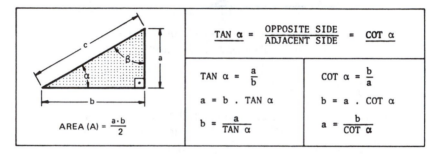

Figure 2.14   Tangent and cotangent functions.

2.6.5 Angular Relationship Between Trigonometric Functions

If $a = c \cdot \sin \alpha$ and $b = c \cdot \cos \alpha$ from 2.6.3 and $\tan \alpha = \dfrac{a}{b}$ from 2.6.4 then by substitution

$$\tan \alpha = \frac{c \cdot \sin \alpha}{c \cdot \cos \alpha}$$

or

$$\boxed{\tan \alpha = \frac{\sin \alpha}{\cos \alpha}}$$

, $\sin \alpha = \tan \alpha \cdot \cos \alpha$, and $\cos \alpha = \dfrac{\sin \alpha}{\tan \alpha}$

Similarly

$$\tan \alpha = \frac{a}{b} \text{ and } b = \alpha \cdot \cot \alpha \text{ by substitution}$$

$$\tan \alpha = \frac{a}{a \cdot \cot \alpha} \quad \text{therefore} \quad \boxed{\tan \alpha = \frac{1}{\cot \alpha}}$$

$$\text{if } \tan \alpha = \frac{\sin \alpha}{\cos \alpha} = \frac{1}{\cot \alpha}$$

then . . . . $\boxed{\cot \alpha = \dfrac{\cos \alpha}{\sin \alpha}}$

2.7 OBLIQUE TRIANGLES

Sometimes the programmer has to do calculations of angles or sides of triangles that do not have a 90° angle. Some calculation procedures for oblique triangles are shown by Figure 2.15.

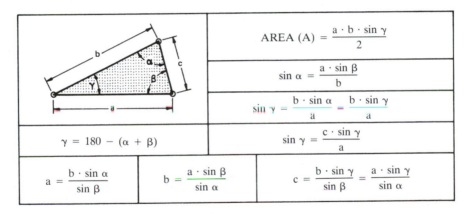

	$\text{AREA (A)} = \dfrac{a \cdot b \cdot \sin \gamma}{2}$
	$\sin \alpha = \dfrac{a \cdot \sin \beta}{b}$
	$\sin \gamma = \dfrac{b \cdot \sin \alpha}{a} = \dfrac{b \cdot \sin \gamma}{a}$
$\gamma = 180 - (\alpha + \beta)$	$\sin \gamma = \dfrac{c \cdot \sin \gamma}{a}$
$a = \dfrac{b \cdot \sin \alpha}{\sin \beta} \quad b = \dfrac{a \cdot \sin \beta}{\sin \alpha}$	$c = \dfrac{b \cdot \sin \gamma}{\sin \beta} = \dfrac{a \cdot \sin \gamma}{\sin \alpha}$

Figure 2.15 Oblique triangles.

EXERCISE

2.7.1. Calculate the three missing elements in oblique triangles using the following data:
(**1**) $b = 2.35$; $B = 39.9°$; $C = 11.8°$ (**2**) $a = 2.22$; $A = 13.4°$; $C = 77.2°$

2.8 ANALYTIC GEOMETRY

Analytic geometry is the science that deals with the graphical representation of an equation. We are mainly interested in introducing the reader to points, lines, and circles, their intersections and relationships in a coordinate system. The reason behind this interest is that outside and inside contours of most parts machined on CNC equipment can be defined in terms of lines and circles. Programmers interested in studying more complex curves such as parabola, ellipse, hyperbola, and others, as well as three-dimensional analytic geometry will find specialized texts dealing exclusively with this topic.

2.8.1 Equation of a Straight Line

A line may be defined through its Y-intercept (the point at which it intersects the Y-axis) and its slope in relation to the positive X axis. See Figure 2.16. The slope-intercept equation can be written as:

$$y = m \cdot x + b$$

Where

$$m = \tan \alpha \text{ and}$$

$$b = 8$$

Example

1. Find the "y" dimensions for

$$X = 6.0 \quad \text{and} \quad X = 8.0$$

$$\text{if } \alpha = 30° \quad \text{and} \quad b = 8$$

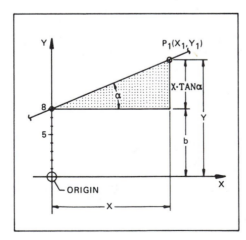

Figure 2.16 Linear graph.

Solution   Both points are located on the above line. Their coordinate axes must therefore satisfy the requirements of the equation.

Tan $30° = 0.57735$; this is the slope m.

Inserting the values of m and b, we obtain the equation of the line:

$$y = 0.57735 \cdot x + 8$$

$$\text{For} \quad x = 6 \quad y = 0.57735 \cdot 6 + 8 = 11.4641 \quad \text{and}$$

$$\text{For} \quad x = 8 \quad y = 0.57735 \cdot 8 + 8 = 12.6188$$

If, on the other hand, we know the coordinates (x_1, y_1) of a point P_1 and the slope of the line, its equation can be obtained from the following formula:

$$\boxed{y = y_1 + m \cdot (x - x_1)}$$

2. Find the slope-intercept form of the equation of a line defined by an angle $\alpha = 30°$ and passing through a point P_1 of coordinates $x_1 = 6.0$ and $y_1 = 11.4641$.

Solution

$$\tan 30° = 0.57735 = m$$

$$y = 11.4641 + 0.57735 \cdot (x - 6.0)$$

$$y = 11.4641 + 0.57735 \cdot x - 3.4641 = 0.57735 x + 8.0$$

$$y = 0.57735 \cdot x + 8.0$$

which is the slope intercept equation used at the start of this paragraph. Using the above equations, any y-coordinate can be found in terms of its x-coordinate.

2.8.2 Equation of a Circle

If the center of a circle is at the origin of the coordinate system (see Figure 2.17), the equation of the circle is:

$$\boxed{r^2 = x^2 + y^2}$$

If the center of the circle is not in the origin of the coordinate system, but located in a point $Q\ (X_Q, Y_Q)$, the equation of the circle will be:

$$\boxed{r^2 = (x - x_Q)^2 + (y - y_Q)^2}$$

Applying these basic principles, the programmer will be able to calculate the intersection point coordinates for line-line, line-circle, and circle-circle relationships.

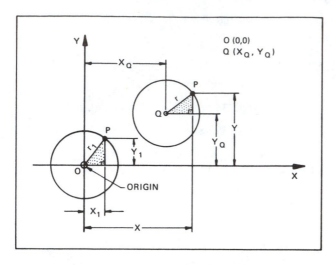

Figure 2.17   Circular graph.

2.8.3 Intersection of Two Lines

Example

Find the intersection point of the following two lines, given in slope-intercept form:

$$\text{LN1} \quad Y = 0.5x + 3.25$$

$$\text{LN2} \quad Y = -2.3x + 7$$

If m_1 and m_2 are the slopes of the two lines, and b_1 and b_2 their respective y-intercepts, the coordinates x_p and y_p of the intersection point PT1 (see Figure 2.18) can be calculated as follows:

$$x_p = \frac{b_2 - b_1}{m_1 - m_2}$$

$$y_p = m_1 \frac{b_2 - b_1}{m_1 - m_2} + b_1$$

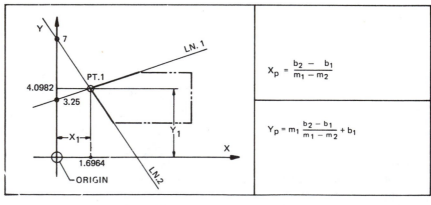

Figure 2.18   Intersection of two lines.

Solution

$$x_p = \frac{b_2 - b_1}{m_1 - m_2} = \frac{7 - 3.25}{0.5 - (-2.3)} = \frac{3.75}{2.8} = 1.6964$$

Substitute this value of x_p into the equation of LN1 as follows:

$$y_p = 0.5 \cdot 1.6964 + 3.25 = 4.0982$$

2.8.4 Intersection of a Circle and a Line

The coordinates of both points of intersection between the line and circle must satisfy both their equations (see Figure 2.19). Therefore, the equation of the line is equal to the equation of the circle as follows:

$$Y_p = m \cdot X_p + b = Y_Q \pm \sqrt{r^2 - (X - X_Q)^2}$$

NOTE: $Y_Q \pm \sqrt{r^2 - (X - X_Q)^2}$ *was derived from the equation of the circle as shown:*

$$(Y - Y_Q)^2 = r^2 - (X - X_Q)^2$$

$$Y - Y_Q = \pm\sqrt{r^2 - (X - X_Q)^2}$$

$$Y = Y_Q \pm \sqrt{r^2 - (X - X_Q)^2}$$

Example

Find the coordinates (X_1, X_2, Y_1, Y_2) of the intersection points (PT1, PT2) from the system of equations of:

$$\text{Line (LN1)} \quad y = x + 2$$

$$\text{Circle (CIR1)} \quad y = 4 \pm \sqrt{2^2 - (x - 4)^2}$$

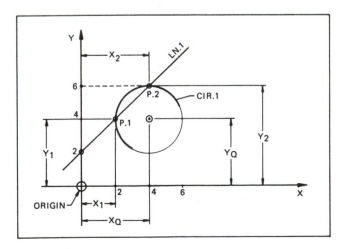

Figure 2.19 Intersection of a line and circle.

Solution

$$y = x + 2 = 4 \pm \sqrt{2^2 - (x - 4)^2}$$

$$(x + 2 - 4)^2 = 4 - (x^2 - 8x + 16)$$

$$(x + 2 - 4)^2 = -x^2 + 8x - 12$$

$$x^2 - 6x + 8 = 0$$

or

$$x = \frac{6 + \sqrt{36 - 32}}{2} = 3 \pm 1$$

$$x_1 = 3 + 1 = 4, \quad x_2 = 3 - 1 = 2$$

$$y_1 = 4 + 2 = 6, \quad y_2 = 2 + 2 = 4$$

and the answers are PT1 (4, 6) and PT2 (2, 4).

2.8.5 Intersection of Two Circles

Example

Find the intersection points of the following two circles:
Circle 1 of equation $y = \sqrt{r^2 - x^2}$ for $r = 4$ and
Circle 2 of equation $y = y_Q \pm \sqrt{r^2 - (x - x_Q)^2}$ for $x_Q = y_Q = 3$
                                                          and $r = 3$

Solution   Since the intersection points are common for both circles, we can equate the two equations as:

$$\sqrt{r^2 - x^2} = Y_Q \pm \sqrt{r^2 - (x - x_Q)^2}$$

Substituting:

$$\sqrt{4^2 - x^2} = 3 \pm \sqrt{3^2 - (x - 3)^2}$$

Squaring both sides:

$$16 - x^2 = 9 \pm 6\sqrt{-x^2 + 6x} + (-x^2 + 6x)$$

Solving the equation:

$$\left(\frac{7}{6} - x\right)^2 = -x^2 + 6x \text{ which becomes } x^2 - 4.16x + 0.68 = 0$$

Solving the quadratic equation for x, we obtain:

$$x = \frac{4.16 \pm \sqrt{4.16^2 - 2.72}}{2} = 2.08 \pm 1.91$$

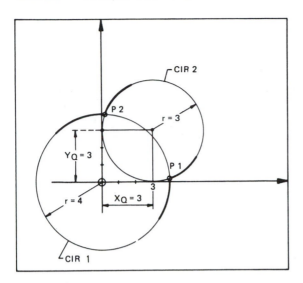

Figure 2.20 Intersection of two circles.

And the coordinates are:

$$x_1 = 2.08 + 1.91 = 3.99 \quad \text{and} \quad y_1 = \sqrt{16 - 3.99^2} = 0.28$$

$$x_2 = 2.08 - 1.91 = 0.17 \qquad y_2 = \sqrt{16 - 0.17^2} = 3.99$$

The two intersection points (see Figure 2.20) will therefore be:

$$\text{P1 } (3.99, 0.28) \quad \text{and} \quad \text{P2 } (0.17, 3.99)$$

EXERCISES

2.8.1. Find the intersection point of the following two lines:
   (1) $y = 2x + 3$; $y = \frac{1}{3}(x - 4)$
   (2) $y = 2x - 3$; $y = -x + 3$
   (3) $y = 0.6x - 0.2$; $y = -\frac{1}{3}x + 1.5$

2.8.2. Find the intersection points of the following line and circle:
   (1) L1 : $y = x$;   C1 : $y = \pm \sqrt{25 - x^2}$
   (2) L1 : $y = 2$;   C1 : $y = 2 \pm \sqrt{25 - (x - 1)^2}$
   (3) L1 : $y = 0$;   C1 : $y = -2 \pm \sqrt{50 - (x + 4)^2)}$

2.8.3. Find the intersection points of the following two circles:
   $y = \pm \sqrt{16 - x^2}$; $y = \pm \sqrt{25 - (x - 3)^2}$

2.9 TRIGONOMETRIC FORMULAS

Formulas discussed in this section will be most useful to the programmer for calculating cutter centerlines for milling applications and TNR centerline paths for turning applications. In milling, "r_c" will be used to identify end mill radius, while in turning, the same "r_c" will represent the tool tip radius. The formulas and sample calculations are given in X-Y coordinates for milling. The reader should have no difficulty in applying these formulas to turning in X-Z coordinates by substituting "Z" for "X" and "X" for "Y" dimensions. See Figure 2.21a and b.

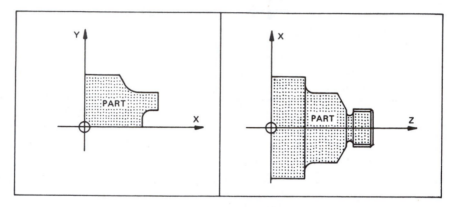

Figure 2.21   (a) Milling. (b) Turning.

2.9.1 Cutter Centerline Intersection Point of a Line Parallel with the X-Axis (Z for Lathe) and a Line at an Angle Measured from the First Line

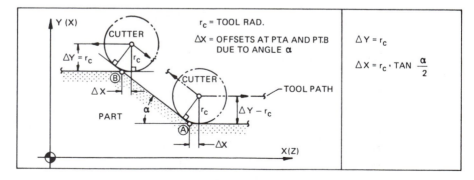

Figure 2.22   Line parallel with the X-axis.

2.9.2 Cutter Centerline Intersection Point of a Line Parallel to the Y-Axis (X for Lathe) and a Line at an Angle Measured from the X-Axis

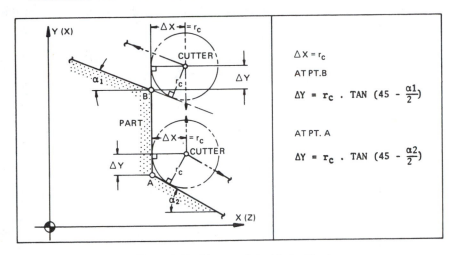

Figure 2.23   Line parallel with the Y-axis.

2.9.3 Cutter Centerline Intersection Point of Two Lines

Neither line is parallel with the primary (X-Y) axes of the part or machine coordinate systems. See Figure 2.24.

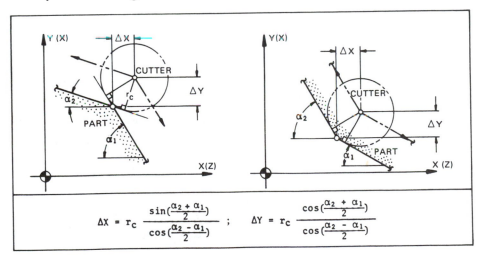

Figure 2.24   Lines not parallel with primary axes.

2.9.4 Cutter Centerline Intersection Point of a Line and a Circle

The line tangent to the circle, not parallel to either X- or Y-axis of the coordinate system. See Figure 2.25.

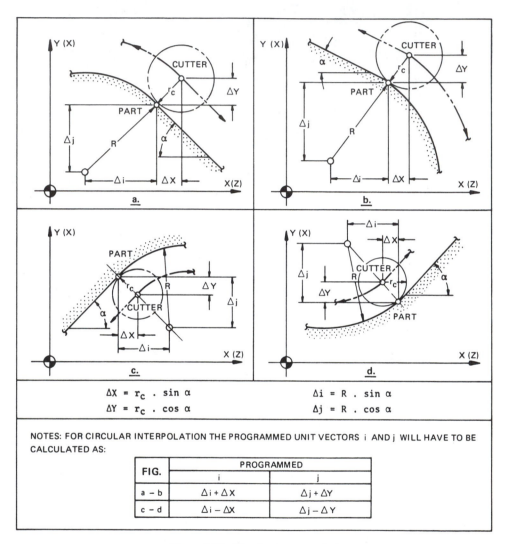

$$\Delta X = r_c \cdot \sin \alpha \qquad \Delta i = R \cdot \sin \alpha$$
$$\Delta Y = r_c \cdot \cos \alpha \qquad \Delta j = R \cdot \cos \alpha$$

NOTES: FOR CIRCULAR INTERPOLATION THE PROGRAMMED UNIT VECTORS i AND j WILL HAVE TO BE CALCULATED AS:

FIG.	PROGRAMMED	
	i	j
a – b	$\Delta i + \Delta X$	$\Delta j + \Delta Y$
c – d	$\Delta i - \Delta X$	$\Delta j - \Delta Y$

Figure 2.25 Line tangent to a circle.

2.9.5 Cutter Centerline Intersection Point of a Circle and a Line:

The lines parallel to the X- or Y-axis, intersecting a circle. See Figure 2.26.

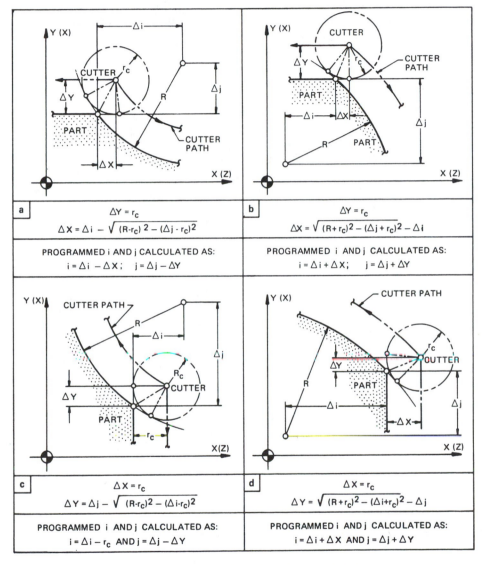

a
$$\Delta Y = r_c$$
$$\Delta X = \Delta i - \sqrt{(R-r_c)^2 - (\Delta j \cdot r_c)^2}$$

PROGRAMMED i AND j CALCULATED AS:
$$i = \Delta i - \Delta X; \quad j = \Delta j - \Delta Y$$

b
$$\Delta Y = r_c$$
$$\Delta X = \sqrt{(R+r_c)^2 - (\Delta j + r_c)^2} - \Delta i$$

PROGRAMMED i AND j CALCULATED AS:
$$i = \Delta i + \Delta X; \quad j = \Delta j + \Delta Y$$

c
$$\Delta X = r_c$$
$$\Delta Y = \Delta j - \sqrt{(R-r_c)^2 - (\Delta i - r_c)^2}$$

PROGRAMMED i AND j CALCULATED AS:
$$i = \Delta i - r_c \ \text{AND}\ j = \Delta j - \Delta Y$$

d
$$\Delta X = r_c$$
$$\Delta Y = \sqrt{(R+r_c)^2 - (\Delta i + r_c)^2} - \Delta j$$

PROGRAMMED i AND j CALCULATED AS:
$$i = \Delta i + \Delta X \ \text{AND}\ j = \Delta j + \Delta Y$$

Figure 2.26   Intersection points of lines and circles.

2.9.6 Cutter Intersection Point of a Line Tangent to Two Circles

A typical turning application is illustrated by Figure 2.27.

These formulas can only be useful to the programmer if a working knowledge is gained by solving numerous problems. Readers wishing to expand their mathematical knowledge beyond the scope of this chapter should refer to specialized manuals of analytical geometry.

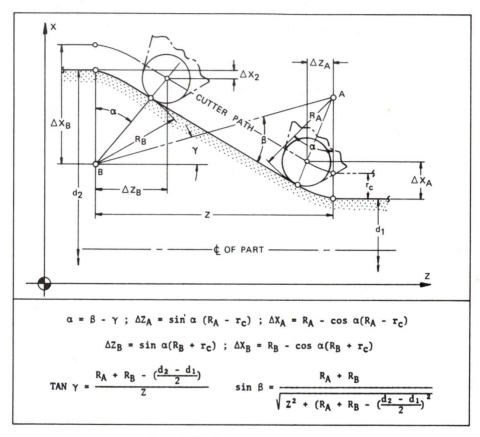

$$\alpha = \beta - \gamma \;\; ; \;\; \Delta Z_A = \sin \alpha \, (R_A - r_c) \;\; ; \;\; \Delta X_A = R_A - \cos \alpha (R_A - r_c)$$

$$\Delta Z_B = \sin \alpha (R_B + r_c) \;\; ; \;\; \Delta X_B = R_B - \cos \alpha (R_B + r_c)$$

$$\text{TAN } \gamma = \frac{R_A + R_B - \left(\dfrac{d_2 - d_1}{2}\right)}{Z} \qquad\qquad \sin \beta = \frac{R_A + R_B}{\sqrt{Z^2 + \left(R_A + R_B - \left(\dfrac{d_2 - d_1}{2}\right)\right)^2}}$$

Figure 2.27   Line tangent to two circles.

2.10 UNIT VECTORS AND DIRECTION COSINES

In a number of instances, the tool-part orientation has to be expressed mathematically for programming purposes. This is achieved by using *unit vectors* and their axis projections, the *direction cosines*. To facilitate the understanding of the

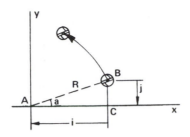

Figure 2.28 Circular interpolation.

above terms, we shall first consider a standard case of circular interpolation, counterclockwise, in the XY plane. See Figure 2.28.

Circular interpolation is programming of the tool motion along a portion of a circular path. Dimensions i and j are the coordinates of the current location of the cutter, or "start point," on the arc shown, measured from the center of the arc. Using Pythagora's theorem in triangle ABC, the radius is calculated as follows:

$$R = \sqrt{i^2 + j^2}$$

The tool motion shown has a *direction*, shown by the arrow on the arc, and a *magnitude*, given by the value of the radius. The values i and j can be seen as the *projections* of the radius R on the axes X, respectively, Y. Their magnitude and direction locates the starting position of the tool point with respect to the center of arc.

If we assume the resulting dimension of the radius R (the "magnitude") to be 1, we have just introduced the *unit vector* (the radius R of magnitude 1) and the *direction cosines i and j*.

The *unit vector* is therefore a "radius" or a "line," starting out from the origin of the coordinate system. Its length is 1 inch (or 1 mm). It is the value 1—hence "unit" vector—which is the critical factor. The measuring system is not relevant in this case. See Figure 2.29.

The magnitude and the direction of the circular interpolation radius R had been determined by its projections i and j. The role of the unit vector is to quantify a direction. To concentrate on the direction, the magnitude is removed from the

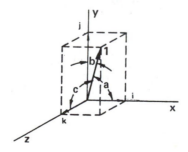

Figure 2.29 Unit vector.

TABLE 2.6

AXES	X	Y	Z
Direction cosines	i	j	k
Tool parallel to X-axis	1	0	0
Tool parallel to Y-axis	0	1	0
Tool parallel to Z-axis	0	0	1
Tool @ 45° in XY-plane	0.707	0.707	0
Tool @ 45° in YZ-plane	0	0.707	0.707
Tool @ 45° in XZ-plane	0.707	0	0.707
Tool equidistant from positive X, Y, and Z axes	0.577	0.577	0.577
Tool @ 30° with respect to positive X axis in XY-plane	0.866	0.500	0

picture by being made equal to one (hence the "unit" in the unit vector). The unit vector expresses the direction by the values of its projections on the three axes, X, Y, and Z. These projections, labelled respectively as i, j, and k, are called *"direction cosines."*

Table 2.6 provides values for a few situations. Negative values indicate the opposite direction. The values have been rounded off to three decimal places for simplicity. They can be calculated accurately if required by observing that

$$\sqrt{i^2 + j^2 + k^2} = 1$$

$$\cos a = \frac{i}{\sqrt{i^2 + j^2 + k^2}} = \frac{i}{1} = i$$

$$\cos b = \frac{j}{\sqrt{i^2 + j^2 + k^2}} = \frac{j}{1} = j$$

$$\cos c = \frac{k}{\sqrt{i^2 + j^2 + k^2}} = \frac{k}{1} = k$$

2.11 UNDERSTANDING THE FUNDAMENTALS OF INTERPOLATION

Linear interpolation represents the machining of a straight-line path between an initial and a terminal location of a cutting tool. These two locations, given by coordinates (x_1,y_1) for the first point and (x_2,y_2) for the second one, will define a tool path given by the straight-line equation

$$y = ax + b$$

Plugging in the two pairs of points, we obtain two equations,

$$y_1 = ax_1 + b_1$$

$$y_2 = ax_2 + b_2$$

These two equations will yield the two unknowns a and b, with the required ratios of pulses to the X and Y axis position control systems. At the programmer's level, all that is required is a block of information showing the end point of the straight-line path (the start point is known as it represents the end point of the previous tool motion).

Circular interpolation has been developed as a standard control function, to generate an arc as a continuous curve, due to the high proportion of circular arcs found in machined parts. As later examples in manual programming will show, the block of information required will contain the end point of the arc, information defining the radius size and position, as well as the direction of travel. If the center of the circle is located at point Q of coordinates x_Q and y_Q, and the circle radius has the value r, the general equation of the circle is

$$r^2 = (x - x_Q)^2 + (y - y_Q)^2$$

Solving this equation for y, and inputting very small increments for x, the control will guide the cutter along a circular path well within specified tolerances.

Helical interpolation is a three-dimensional extension of circular interpolation. The block requires a plane statement, the end point in three-dimensional coordinates, the radius, and the direction of travel. The projection of the helix on the plane defined will be a circle of the radius specified.

Parabolic interpolation is a synonym for curve fitting. A circle connected two positions of the tool, a start position and an end position. Its shape was given by the circle radius. The parabolic interpolation can connect three positions, using a parabolic curve. The parabola is usually defined as a set of points, each of which satisfies the condition that its distance from a fixed point (focus) is equal to that from a fixed line (directrix).

A special purpose computer system will look at a number of positions. It will fit a parabola based on positions 1, 2, and 3 and store the data for the 1–2 portion. It will then fit a parabola over points 2, 3, and 4 and store the data for the 2–3 portion, and so on. If the distance between points is sufficiently small, smooth sculptured patterns or surfaces can be produced.

To the same extent, the *parabolic shape* can be replaced by an *elliptic shape*, in which case we have *elliptic interpolation*. There is, however, nothing sacred about the various interpolation alternatives. They usually require the acquisition of special control options, now superseded by special-purpose machine control combinations. Advanced general purpose controls solve the problem by user macros (parametric subroutines). Any curve, no matter how complex, can be machined so long as it can be defined mathematically. This brings to the shop floor the ability to machine complex shapes, which used to require computer-assisted programming based on one of the major computer graphics packages. User macros will be examined in detail in a later chapter.

EXERCISES

2.9.1. In Figure 2.22, p. 36, given $\alpha = 30°$ and $r_c = 0.75$ in., calculate X and Y.

2.9.2. In Figure 2.23 (point A), p. 37, given $\alpha_2 = 25°$ and $r_c = 0.625$ in., calculate ΔX and ΔY.

2.9.3. In Figure 2.24, p. 37, given $\alpha_1 = 45°$, $\alpha_2 = 15°$ and $r_c = 1.00$ in., calculate ΔX and ΔY.

2.9.4. In Figure 2.25a, p. 38, given $\alpha = 40°$ and $r_c = 0.5$ in., calculate ΔX, ΔY, Δi, Δj, i and j

2.9.5. In Figure 2.26a, p. 39, given $\Delta i = 4.0$ in., $\Delta j = 3.0$ in., and $r_c = 1.0$ in., calculate R, ΔX and ΔY.

2.9.6. Calculate the coordinates of points A and B in Figure 2.30.

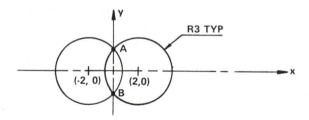

Figure 2.30   Intersection of two circles.

2.9.7. Calculate the coordinates of points A, B, C, and D in Figure 2.31.

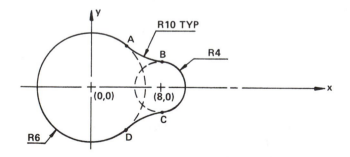

Figure 2.31   Connecting two circular shapes.

2.9.8. Calculate the coordinates of points A, B, C, D, E, and F, as well as the distance AB in Figure 2.32.

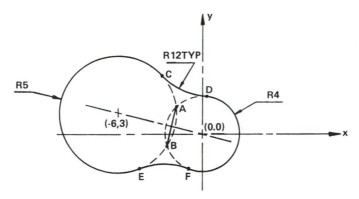

Figure 2.32 Connecting circles and calculating distance.

2.9.9. Calculate the coordinates of points A, B, C, D, E, F, and G in Figure 2.33. Calculate angles M, N, and P. Verify that $M + N + P = 180°$.

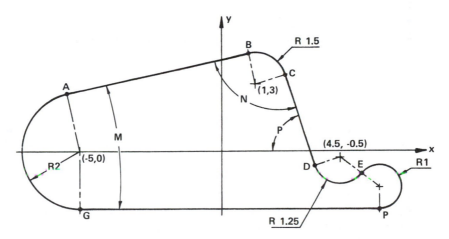

Figure 2.33 Complex shape.

3

Computer Numerical Control Systems

In the previous chapter we discussed how the special-purpose CNC accepts information in the form of punched or magnetic tape codes. This input data must be transformed by the CNC into specific output codes in terms of voltages, or pulses per second (pps). The transformed data, output, is used to drive the motors to position the machine slides to the programmed position. These slides, or table drives, are commonly known as servodrives. The principal function of the CNC is the positioning of the tool or the machine table in accordance with the programmed data. Industry has developed two distinctly different types of drives based on how the CNC system accomplishes positioning. These are open-loop and the closed-loop drive systems.

3.1 OPEN-LOOP SERVODRIVES

An automatic washing machine is a typical example of an open-loop system. It will perform a fixed cycle regardless of the state of cleanliness of its contents. In the open-loop servodrive control system, the power supply level is set to a position for which the desired speed is indicated by the input. Should the load on the machine slide vary, the servomotor speed would also be affected. However, the speed variance could not be sensed automatically because the system lacks feedback. For

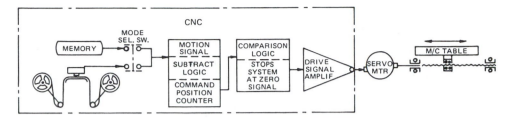

Figure 3.1   Open-loop CNC positioning control.

this reason, open-loop control systems can only be used in applications in which there is no change in load conditions. A typical application of the open-loop control is the NC drilling machine. The function of the servodrive is to position the machine slide or table; therefore, the load condition remains constant. Changes in load due to the weight of the component being machined are taken care of by the design of the machine tool. Manufacturers will normally specify the maximum admissible weight of the parts that can be machined on their product. See Figure 3.1.

In the open-loop servosystem, the motor continues to turn until the absence of power indicates that the programmed location has been attained and the driving mechanism is disengaged. There is no monitoring of this position, and if any movement takes place at this time, its magnitude would normally be unknown. Nevertheless, open-loop control systems have been refined to 0.0001 inch (0.0025 mm) resolution. The systems are reliable, considerably less expensive than closed-loop systems, and their maintenance is far less complicated. Periodic adjustments are required to compensate for wear, as well as deterioration of servodrive components. In summary, this system counts pulses, and it cannot identify discrepancies in position.

3.2 CLOSED-LOOP SERVODRIVES

The closed-loop system used in CNC is characterized by the presence of feedback. The term *feedback* is used to describe the various methods of transmitting positional information on the machine slide motion back to the information command section of the CNC. This information is continuously compared with the programmed slide motion data.

CNC systems use two different feedback principles. The indirect feedback monitors the output of the servomotor, as shown in Figure 3.2a. Although this method is popular with CNC systems, it is not as accurate as direct feedback, which monitors the load condition in the feedback loop, as shown in Figure 3.2b.

The feedback device is commonly known as "transducer," and may be represented by linear or circular electric scales, shaft digitizers, magnetic scales, or synchros. All these different feedback methods have one thing in common. If the system is digital, the feedback device is counting pulses. If it is analog, the feedback

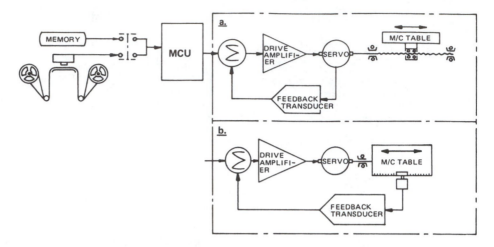

Figure 3.2 Schematic diagram of feedback systems. (a) Indirect feedback. (b) Direct feedback.

is comparing varying voltage levels. A closed-loop system, regardless of the type of feedback device, will constantly try to achieve and maintain a given position by self-correcting to a zero pulse or a voltage null. The indirect feedback system (Figure 3.2a) compares the command position signal with the drive signal of the servomotor. This system is unable to sense backlash or leadscrew windup due to varying loads. The direct feedback, with its drive signal originated by the table, is the preferred system because it monitors the actual position of the table on which the part is mounted. The direct feedback system is also called "positional" feedback system. It is more accurate; however, its implementation costs are higher. While the subject of feedback and servodrives is a science in its own right, a description of a rotary-type transducer used in a digital system (see Figure 3.3) may illustrate a small portion of a control system.

The rotary disk is attached to the moving part of the machine table. The

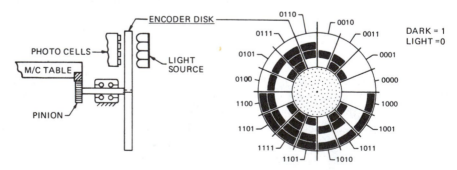

Figure 3.3 Rotary encoder for digital transducer.

TABLE 3.1 DECIMAL-BINARY-GRAY CODES

Decimal	Binary	Gray	Decimal	Binary	Gray
0	0000	0000	8	1000	1100
1	0001	0001	9	1001	1101
2	0010	0011	10	1010	1111
3	0011	0010	11	1011	1101
4	0100	0110	12	1100	1010
5	0101	0111	13	1101	1011
6	0110	0101	14	1110	1001
7	0111	0100	15	1111	1000

pinion (disk) is rotated as the table movement takes place, and each light interruption from the cells creates an electric pulse. If this pulse is the minimum programmable increment, we have a direct one-to-one feedback to the drive command. The encoder is a rotary disk with clear and darkened areas, and it is designed so that for each fixed fraction of a degree, another binary combination is encoded. Figure 3.3 shows a 4-bit encoder using "Gray-code," a cyclic code generating one change at a time. See Table 3.1. Binary encoders used on earlier systems produced erroneous readings because of ambiguity, or overlap, in reading a code position. The problem was compounded by higher accuracy requirements when 15- or 16-bit resolution was required with a small disk.

A four-resolution segmentation is calculated as follows:

$$360/2^4 = 360/16 = 22.5^0$$

In the previous calculation, the power of 2, i.e., 4, represents the number of rings. The disk, when rotated clockwise, will read the Gray codes as marked. These codes can easily be converted to binary using a simple single flip-flop.

Usually optical disks containing 2^{10} to 2^{15} bits are used. For $2^{10} = 1024$, the segmentation would be $360/2^{10} = 360/1024 = 0.3515^0$. This requires instrumentation manufacturing techniques for production, and, as a result, the disks become quite expensive.

3.3 VELOCITY FEEDBACK

The positional feedback previously discussed can be implemented to a very high degree of accuracy. However, the system may not provide the required path or surface finish accuracy because no time constraints were provided to reach the programmed or final position. CNC systems used for contouring must have a velocity feedback as well, in order to produce linear, circular, and/or parabolic interpolation, at the same time as acceleration and deceleration velocities.

Feedback is normally provided by an AC or DC tachometer coupled to the servomotor. The feedback of the tachometer is used to modify the positional feedback. The feedback schematics shown in Figure 3.1 and Figure 3.2 do not incorporate a velocity loop. Most actual CNC position feedback loops provide some kind of velocity control inside the positioning loop, even if it is not specifically expressed.

The importance of the velocity loop can best be described by the example discussed below.

Example

Calculate the displacement inaccuracy of a CNC pulse drive system without a velocity feedback for the following motion:

N019 G01 X82.55 Y44.45 F100

(in metric), or

N019 G01 X3.25 Y1.75 F3.94

(in inches), where
N019 is the sequence number of the tape block
G01 is the preparatory code for linear interpolation
X82.55 is the programmed displacement or address along X-axis
Y44.45 is the programmed displacement or address along Y-axis
F100 is the tool velocity, in mm/min in the metric line.
The following assumptions are made:

- The smallest programmable increment is 0.002 mm (0.0001 inches)
- The CNC has pulse servodrives
- To produce the tool path illustrated in Figure 3.4, both slides must start and stop simultaneously.

Solution

1. Calculate the number of pulses required for both the X82.55-mm and Y44.45-mm displacements.

$$\frac{82.55}{0.002} = 41,275 \text{ pulses}$$

$$\frac{44.45}{0.002} = 22,225 \text{ pulses}$$

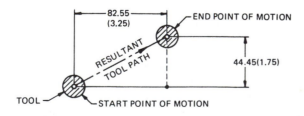

Figure 3.4 Tool path.

Since both slides must start and stop simultaneously, these calculated pulses corresponding to the total displacement must be distributed on a different time scale to produce the resultant tool path.

2. Calculate the number of pps required to generate the tool path.

- X time interval

$$\frac{82.55 \text{ mm}}{100 \text{ mm/min}} = 0.8255 \text{ minutes}$$

$$\frac{41,275 \text{ pulses}}{0.8255 \text{ min.}} = 50,000 \text{ ppm}$$

$$= 833.333 \text{ pps}$$

- Y time interval

This is established as a function of the X-displacement, in a direct ratio of the two programmed values for X and Y.

$$\frac{44.45}{82.55} = 0.538462$$

$$0.538462 \cdot 833.333 = 448.718 \text{ pps}$$

3. Calculate the error.

The CNC cannot generate fractional pulses. Therefore, the inaccuracy of the system can be calculated as follows:

- X-error

0.8255 minutes · 60 = 49.53 seconds motion time
49.53 · 833 pps = 41,258 pulses (compare with 41,275)
41,258 pulses · 0.002 mm = 82.5169 mm
The error = 82.55 − 82.5169 = 0.033 mm or 0.0012 inches.

- Y-error

49.53 · 448 pps = 22,189 pulses (compare with 22,225)
22,189 pulses · 0.002 mm = 44.3788 mm
The error = 44.45 − 44.3788 = 0.071 mm or 0.0027 inches.

In the case of the velocity feedback, the errors can be substantially reduced by applying a larger time scale.

The DC drive systems can work with fractional voltages; therefore, this error would not occur.

Current CNC servosystems are refined to a degree that manufacturers can guarantee 0.002-mm accuracy without difficulty.

The control for circular or parabolic interpolation is a great deal more intricate. The pps must constantly be varied for both axes in order to generate an arc. Readers who wish to study the control aspects of circular or parabolic interpolation should have no difficulty finding texts on the topic of servodrive design.

EXERCISE

3.3.1. Calculate the displacement inaccuracy of a CNC pulse drive system having no velocity feedback, for the following motions:

N069 G01 X3.756 Y6.925 Z1.25 F6.0

(in inches), and

N070 G01 X62.05 Y17.372 Z22.111 F221

(in mm).

For each axis, calculate the following data in inches and metric separately:
- Number of pulses required to carry out programmed motion
- The time corresponding to each set of pulses
- The respective pulse rate in pulses/min (ppm) and pulses/second (pps)
- The error, or inaccuracy

3.4 POINT-TO-POINT POSITIONING CONTROL

The principal function of the point-to-point positioning control is to position the tool from one point to another within a coordinate system; therefore, the control is most often referred to as a point-to-point NC system. The positioning may be linear in the X-Y plane, or linear and rotary if the machine has a rotary table. Each tool axis is controlled independently; therefore, the programmed motion may be simultaneous or sequential, but always in rapid traverse. Machining can only take place after positioning is completed. The most common applications of the point-to-point control are in drilling, boring, tapping, riveting, pipe bending, and sheet metal punching.

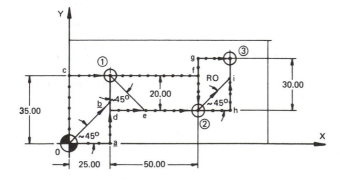

Figure 3.5 Tool path of point-to-point control.

TABLE 3.2 PATHS OF THREE DRILLED HOLES

Programmed	Tool Path		Motion	
	Sequential	*Simultaneous*	*From*	*To*
X25.00 Y35.00	0-a-1 or 0-c-1	0-b-1	0	1
X50.00 Y-20.00	1-d-2 or 1-f-2	1-e-2	1	2
X20.00 Y30.00	2-h-3 or 2-g-3	2-i-3	2	3

In addition to positioning, this system is capable of controlling auxiliary functions such as tool change, spindle and coolant on and off, part or fixture clamping, indexing, etc. These on/off type of relay functions do not require intricate control logic, nor do they have any velocity (feed) control.

The tool path of point-to-point control is illustrated in Figure 3.5. If the positioning is sequential, the system will move in one axis at a time (as illustrated by the dotted lines). If the positioning is simultaneous, both axes start to move at the same time. Assuming that both drives have the same speed, the tool path will be approximately 45° to the point where the lesser of the two dimensions was completed (as illustrated by solid lines). From this intermediary point to the final programmed point, the motion will be parallel with the primary axis of the system. The paths described in Table 3.2 are an additional illustration of the drilling of the three holes shown in Figure 3.5.

Most point-to-point NC systems are built with open-loop drives.

3.5 STRAIGHT-CUT POSITIONING SYSTEMS

The straight-cut positioning system provides a limited degree of control during the positioning of the tool from one point to another. Most of the straight-cut systems are fitted with manually adjustable feed control. This feed control is shared by all the programmable axes of the NC machine, which allows the system to perform milling, in addition to the drillinglike operations outlined in the previous paragraph. Because of this shared feed control feature, the system can also perform milling operations at 45° to the primary axes of the machine.

Realizing this limitation, the programmer can reduce the programmed steps to small enough increments to mill any straight-line pattern and produce the required surface finish and accuracy for most industrial applications. The accuracy this system can produce would most certainly be sufficient for any rough milling application.

The tool path of the straight-cut system is shown in Figure 3.6 and by the following example:

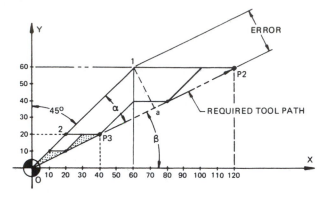

Figure 3.6 Tool path of straight-cut system.

Example

The coordinates of point P2 are X120.00 mm and Y60.00 mm.

Programming these dimensions would result in a tool path from 0 to 1 to P2, with the following error:

$$\text{Error} = \text{distance } 1a = 01 \cdot \sin \alpha = \sqrt{x_1^2 + y_1^2} \cdot \sin \alpha = \sqrt{2(y_1)^2} \cdot \sin \alpha$$

$$\text{since } x_1 = y_1$$

$$\text{and } \alpha = 45° - \beta = 45° - 26.565° = 18.435°$$

The value of β was obtained from the X-Y coordinates of point P2.

Continuing the calculation.

$$\text{Error} = \sqrt{2 \cdot (y)^2} \cdot \sin 18.435° = \sqrt{2 \cdot (60)^2} \cdot \sin 18.435° =$$

26.8328 mm

This value of the error is obviously too large.

Reducing the programmed increments to X40.00 mm and Y20.00 mm would result in a tool path of 0 to 2 to P3, and the error would be:

$$\text{Error} = \sqrt{2 \cdot (y)^2} \cdot \sin \alpha = \sqrt{2 \cdot (20)^2} \cdot \sin 18.435° = 8.944 \text{ mm}.$$

This value is proportionally less, but still far too large.

Using this method, we can reduce the programmed increment to steps small enough to yield the desired accuracy.

If the programmed steps are X0.5 mm and Y0.25 mm, the error would be:

$$\text{Error} = \sqrt{2 \cdot (25)^2} \cdot \sin 18.435° = 0.039 \text{ mm}.$$

This value is sufficient for rough milling and for some finishing applications. The steps could be further reduced; however, the programming time and the length of the tape would become prohibitive.

Straight-cut positioning controls can be built with open- or closed-loop feedback drives, with programmable accuracies of 0.002 mm (0.0001 inch). These systems are most often built with programmable spindle speed and with various preparatory (G) codes.

3.6 CONTOURING, OR CONTINUOUS PATH CNC SYSTEMS

The contouring system is the state-of-the-art, high-technology, most versatile and intricate of the CNC devices.

It generates a continuously controlled tool path, by intepolating intermediate points or coordinates. By "interpolating" we mean the capability of computing the points of the path.

All CNC contouring systems have the ability to perform linear interpolation. This feature is a computer subroutine, permanently recorded in the CNC computer under a G01 preparatory code. G01, programmed with X-Y or X-Y-Z dimensions, is an instruction to interpolate a tool path, in its shortest distance between two points.

The best way to illustrate the usefulness of this feature is to compare the programming effort difference between contouring and the previously discussed straight-cut system.

In order to program the tool path from 0 to P1 in Figure 3.6, we would require a single tape block on a contouring system, as shown below:

N100 G01 X120.00 Y60.00 F. . .

This line will produce a tool path to 0.002-mm tolerance.

Using the straight-cut system, the program shown below will require 240 tape blocks and will only yield a tolerance of 0.039 mm.

N1 . . . . . X0.50 Y0.25	from zero to X0.50, Y0.25
N2 . . . . . X0.50 Y0.25	to X1.00, Y0.50
N3 . . . . . X0.50 Y0.25	to X1.50, Y0.75
. . . . .	
. . . . .	
. . . . .	
N240 . . . X0.50 Y0.25	to X120.00, Y60.00

Most contouring systems also have a circular or parabolic interpolation feature. The programming of circular arcs is also done in one tape block; however, in addition to the appropriate preparatory (G) code, additional dimensions must be programmed to identify the relationship between the start point and the arc center.

Most contouring systems have programming capabilities of four or five axes. A large percentage of contouring CNC systems are built with positioning and velocity feedback drives.

3.7 ADVANCED CNCs

CNC builders have taken full advantage of the capabilities of the microprocessors used in the modern controllers. The proliferation of units and features blurs the distinctions that used to exist among CNCs. Thus, for instance, the Allen-Bradley CNC mentioned above can be listed here as well, together with units from well-known suppliers such as Cincinnati Milacron, Giddings & Lewis, Fanuc, General Numerics, GE Fanuc, Kearney & Trecker, Yasnac, Mitsubishi, and many more.

Each manufacturer groups and describes the features of the equipment it produces in different ways. The list below will attempt to provide a generic composite of some of the more advanced features of a number of CNCs:

- Color graphics CRT display—size and resolution
- Memory size—standard and optional
- Tool life management—with automatic cutter compensation of spare tool.
- Resident diagnostics.
- Environmental features—temperature, humidity, power, grounding, sealing, air-conditioning, internal compartment separation.

The latest series of CNC systems have extensive new features, but their most meaningful contribution is in the way these features have been integrated.

Some of the more significant controllers will be examined, in no particular order. It should be noted that some are produced by long-standing machine-tool manufacturers, while others come from companies that are primarily electronic, electrical, automation, or industrial control specialists.

Cincinnati Milacron. Cincinnati Milacron is primarily a machine-tool manufacturer, producing machine-tools, robots, and their controls. Among others, it has a full range of CNC controllers, such as 850 TC CNC series for turning, 850 MC and 950 MC for machining centers, 700 G CNC for grinding applications, as well as robotics controllers and software packages for arc and spot welding, materials handling, plastic injection, and blow molding and extruders. The use of CNC systems has been extended to waterjet cutting machines for control of not only the tool path but the entire technological process including a 60,000 psi waterjet processing system. Other areas are laser cutting of composites, engraving, trim, heat treatment, coordinate measuring machines and electrical discharge machines (EDM), all well-established areas where CNC proved itself to be of primary importance.

Two of the most recent applications are in Flexible Manufacturing Systems (FMS) and Composite Tape Laying (CTL). The FMS uses CNC equipment enhanced with special or added communications and material handling capabilities, as well as some sort of process inspection. CTL is a new technology, and the special purpose 975C CNC system can also handle other fiber placement machinery applications. Up to ten axes may be used to make automated layups of graphite/

epoxy, graphite/bismaleimide (BMI), and other composite pregregs. It is anticipated that the use of thermoset materials will be accelerated by the requirements of the aircraft and aerospace industries.

Cincinnati 850 Series. Cincinnati Milacron Acramatic 850 MC was designed for high-performance machining centers. It provides control for three linear axes plus an optional fourth rotary axis. Some of its more advanced features are:

- Simultaneous machining and programming, with a true "foreground/background" feature, provided by the new family of multitasking controllers.
- The use of custom macros.
- Resident diagnostics with a library of hundreds of errors and malfunctions.
- Multiple display pages providing more information for the operator and greatly reducing the misinterpretation of messages.
- Menu driven multicolor interactive display.
- New tool data management allowing the user to store information such as tool-length, diameter, flutes, etc., as well as the location of over 150 tools. The system also allows sharing the tools among various jobs, keeps track of tool life values, and warns the operator when a tool should be replaced.
- Custom macros for contouring, pocketing, face milling, or hole-pattern machining. The hole-pattern subprogram allows the user to define up to five processes, such as centerdrill, drill, ream, tap, and chamfer. It also provides many other time-saving features for the programmer when writing part programs for rough, rough and finish, rough leave stock, and finish.
- Using multiple G and M codes in the same block.
- Built-in calculator, including trigonometric functions, for quick and easy programming.
- Dry-run graphics, coordinate system rotation, workpiece alignment mode, etc.
- Process interruption no longer presents major problems due to the feature of "automatic return to cut."
- Offsets are expanded from tool to fixture and pallets.
- Fillet/radius feature in menu allows for easy surface blending.
- Use of robots in conjunction with probes for settings or in-process gaging.
- Operator messages or alerts, color coded to emphasize the severity of the problem.
- Record-keeping feature, with the ability to record 256 alerts and up to 995 start-ups, information that can be accessed at any time during the machining cycle.

Cincinnati 950 Series. The 950 Controller is a true state-of-the-art five-axis CNC System, and the integration of its capabilities point to the requirements of the factory of the future. The high-performance Machine Mechanism Control allows it to follow instructions to give commands to other equipment or devices, such as robots or probes for setups or in-process gaging. Its programmable machine interface can expand to 192 K of code and work with up to three independent dedicated processors. Some of the most important features are:

- Touch screen and pop-up windows, activated by user.
- Wordprocessor type editing, allowing for quick search of characters (single, strings, or blocks). The programmer can insert, erase, copy, or define new or old programs.
- 20 MByte hard disk storage, allowing for storage of a large number of library subroutines. This also extends the tool management scope to 200–500 tools, as tables of tool data, location, and wear.
- System efficiency has been increased to include automatic inch-metric capability, high-speed probing, and torque-controlled feed function.
- System security through a password setup with several levels of access privilege.
- System accuracy enhanced through a 20 block look-ahead, allowing for timely acceleration and deceleration in sharp cornering.
- System diagnostics monitor constantly all potential problem areas, particularly the servosystems that can be monitored and adjusted individually on the screen. Paper management information is optional.
- System communications have a high level of management flexibility. The user can interface CNC tape readers, high-speed computers or printers, bar code readers, port counters, etc., including all necessary equipment for management information reports.
- System potential. The open-ended design lends itself to FMS or CIM adaptation, addition of axes, or BCL or ACL implementation.

General Numeric. General Numeric is a sales and service organization for two essentially electronic/electrical companies, Siemens from Germany and Fanuc from Japan. These well-known companies manufacture an almost endless range of products, available under the General Numeric (GN label), as well as their own individual models.

The Fanuc Series 00/100/110/120/150 are the new series of CNCs, integrating three types of control function that require definition:

1. Man-Machine-Control (MMC) refers to the human interface with the control system. Machine-tool builders integrating the control in their system can create software to enhance the performance of the machine, develop and protect conversational mode screens using their choice of programming language, etc.

2. Computerized Numerical Control (CNC) is not a new label. What is new is the fact that it becomes now just one of the integrated parts of the new advanced control.

3. Programmable Machine Control (PMC) refers to the control of auxiliary machine functions such as tool change, spindle control, work handling, etc., by developing, for instance, setup software tailored to a specific machine.

High speed interchange of related information among the above functions allows for the best possible fit between machine-tool, control, and user-production needs.

The Fanuc Series 15 covers a selection of computers developed as Artificial Intelligence CNCs, incorporating the latest hardware advances. Its high speed makes possible the management of high speed and high precision digital servosystems, high-speed PMC functions (unaffected by the fluctuating mechanical loads), intelligent failure diagnostic guidance, etc. It allows programming in Pascal or Ladder language and has modularized inputs/outputs with 1024 input and 1024 output points (I/O). A number of the CNC functions have been enhanced to increase the efficiency of the machining cycle. The system can be further customized by judicious combination of custom macros and MMC, conversational automatic programming, and communications features allowing Local Area Networking (LAN) and connections to the Manufacturing Automation Protocol (MAP).

Mitsubishi Meldas 300.   These have specifications and performance abilities fairly similar to Fanuc. The manufacturer emphasizes modularization and the ability of the machine-tool manufacturer or the user to put together from a very simple CNC to an ultra complex system. Some of the more distinct features are communications, advanced programmable logic controller features, such as automatic measurement, machine diagnostics, and adaptive control. Other special attributes are shop floor graphics, programming, data collection, and many more.

Kearney & Trecker.   This is another machine-tool manufacturer producing a wide product line. One of the more recent systems is the ORION, with its own Gemini Modular D CNC. For customers who have a particular control preference, the system has been designed to accommodate a wide variety of controls, including General Electric/Fanuc or Allen Bradley. Some of the features of the Gemini are modularity, full tool management capabilities, adaptive control of speeds and feeds, hard disk memory for all data, as well as operator assistance through a set of menus and prompts.

Giddings & Lewis.   This company produces a comprehensive line of machine tools as well as the CNC 8000, its new microprocessor-based control. This control has a long list of features, some of which are tape cartridge storage, parametric subroutines, macro subroutines, previous control tape compatibility, 8-axis contouring, ladder diagram logic machine interface, automatic "return to path," automatic in-process gaging, adaptive control, tool retract-reentry, etc.

Bridgeport.   Bridgeport is a well-known machine-tool manufacturer. Its more recent product line is equipped with well-known CNC controllers from Heidenhain and Fanuc.

Allen-Bradley.   This is an industrial control manufacturer which supplies a CNC controller in addition to an extensive line of programmable controls. The series 8200 CNC with 8000-Advanced Technology (AT) Option has features such as a 12-axis capability, an enhanced human interface with soft keys, color CRT, and help files, additional memory and I/O slots, paramacro programming, etc. The combination of additional memory and I/O capacity of the 8000-AT and advanced communications of the 8200 facilitate features such as workcell and FMS integration. Allen-Bradley is also one of the pioneers of networking of its systems to the VistaMAP broadband Local Area Network (LAN).

3.8 BINARY CUTTER LOCATION (BCL)

EIA Standard RS-494 defines BCL as a "32 Bit Binary Cutter Location Exchange (BCL) format for Numerically Controlled Machine Tools." Some of the features of BCL are:

- It defines input command records.
- It outlines the response of the machine tool.
- It orients the machine/control system to the part.
- It implements part program portability among various machine/control systems.
- It provides a standard interface to the various machine tools on the shop floor.
- It eliminates altogether the need for postprocessor purchase, installation, and update/maintenance.

Part-Machine Connection.   The machine/control system has the machine coordinate system defined in a package that can be built in or retrofitted, depending on the vintage of the machine. This software also comprises the relationship between the part and machine coordinate systems, and their reciprocal position must be set at the beginning.

Binary Files.   These represent a collection of records, expressed internally in binary, and represented in the printouts in hex (short for hexadecimal, or base 16 numbering system). The file is called "sequential," as it contains "records" in sequence, one after the other. These records, which may have variable lengths, are defined as groups of 32-bit words (a bit is a binary digit, either 0 or 1). The group starts with a 32-bit header, which could be FEDRAT, COOLNT, etc., or other terms familiar to APT users and easy to understand for nonusers. The sequential file is in reality nothing but the old program, and the records are "program words," grouped in blocks which used to be separated by an EOB (End of Block) code.

Understanding BCL.   A 32-bit number may be considered as comprised of two 16-bit parts. The first 16 bits (the "upper" 16 bits) contain the code for a "major" word, such as FEDRAT or SPINDL. The "lower" 16 bits contain the CL (cutter location) record number in the program. Other 32-bit words may represent "minor" words (such as FLOOD, MIST, etc.), directions, or coordinates. For instance, the Hex number 00E4 E1C0 can be converted to decimal 15 000 000. As the multiplier is 10 000, the conversion has yielded 1500, representing 1500 RPM. These conversions are built in the various computer systems, but for those desiring to try a conversion at least once, using a calculator, here is the method. The above hex number, using 16 digits (from 0 to 15, where beyond 9 we count A for 10, B for 11, etc.), has its digits multiplied by subsequent powers of 16 (hence the "hex"): $16^0 = 1$; $16^1 = 16$; $16^2 = 256$; $16^3 = 4096$; $16^4 = 65536$, etc.

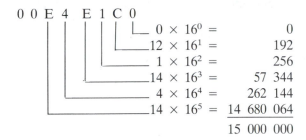

$$
\begin{array}{l}
0 \times 16^0 = 0 \\
12 \times 16^1 = 192 \\
1 \times 16^2 = 256 \\
14 \times 16^3 = 57\;344 \\
4 \times 16^4 = 262\;144 \\
14 \times 16^5 = 14\;680\;064 \\
\hline
15\;000\;000
\end{array}
$$

(NOTE: C in hex corresponds to 12 decimal, and E to 14).

Essentially BCL-formatted data allows the CNC program, once proven, to be used by the machine tool without the need for subsequent postprocessing. Some of the immediate advantages of this process are as follows:

- Standardization of formats for exchanging (transporting) programs between CNC, CAD-CAM, and PLController systems.
- Elimination of need for postprocessors, with its logistic, programming, maintenance and overall cost implications, incompatibility of some programs with some machines, etc.
- Solving the problem of tape-based program storage needs, as the floppy disc stores the equivalent of 1000 ft of tape in a small fraction of the required space.
- Solving the non-stop juggling of machine scheduling, when the machine available at a particular time is not the one for which the next part's program was post-processed. The solutions available to date, all costly, were waiting for the machine to become available, prepare a newly post-processed program for a different machine, or prepare several different programs for each part/ machine possibility.

- Most reasonably recent controllers can be converted to accept BCL data with minimal problems and cost.

A sample of the BCL approach can be illustrated in the following:

RS-274 (present standard)	RS - 494 - BCL
N____ G20 (or G70)	Units/Inches
N____ G21 (or G71)	Units/Metric
N____ M08	Coolant/Flood
N____ F300	Fedrat/30
N____ G1 X − 5.00 Y3.25 Z − 4.75	GOTO/−5,3.25,−4.75

BCL can be obtained as direct output of CAD/CAM. It uses the part coordinate system, and its data "is" the CNC part program. It will output programs in the BCL part program mode (part cartesian coordinate system). This in turn is input to a BCL machine/control system (machine cartesian coordinate system).

Trends.   Work has been done recently on a modification of BCL, ACL, where "B" as in Binary is replaced by "A" as in ASCII.

4

Machining Forces

The programming of feeds and speeds is the responsibility of the part programmer. While the programming accuracies relate to the dimensional accuracy of the work piece, the programing of feeds and speeds relates to the efficiency of the machining process. As a result, they carry equal importance in CNC programming. As in the previous paragraphs, we shall discuss only the most important formulas. The most important consideration in the selection of speed and feed is the available horse-power of the CNC machine tool. Generally, the power required for drilling, milling, or turning is expressed in terms of the average unit power. The average unit power is the power required to remove 1 cubic inch of metal in 1 minute.

4.1 DRILLING

4.1.1 Cutting Speed (Vc)

See Figure 4.1 for illustration of drill area.
At Point A:
    in inches:

$$V_c = \frac{d \cdot \pi \cdot n}{12} \text{ feet per minute (fpm)}$$

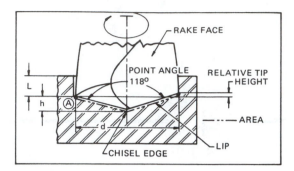

Figure 4.1 Drill area.

in metric:

$$V_c = \frac{d \cdot \pi \cdot n}{1000} \text{ m/min.}$$

Where

$$V_c = \text{Cutting velocity}$$

$$n = \text{Drill rpm}$$

$$d = \text{Drill diameter (in inches or mm)}$$

4.1.2 Rate of Metal Removal (Qd)

In inches:

$$Qd = A \cdot n \cdot F \qquad \text{in in}^3/\text{min}$$

In metric:

$$Qd = \frac{A \cdot n \cdot F}{1000} \qquad \text{in cm}^3/\text{min}$$

Where

$$Qd = \text{Volume of metal removed in in}^3/\text{min or cm}^3/\text{min}$$

$$n = \text{Drill rpm}$$

$$F = \text{Feed in inches or mm per revolution of drill}$$

$$A = \text{Area} = r^2 \cdot \pi = \frac{d^2 \cdot \pi}{4} \text{ in square inches or mm}^2$$

MATERIALS	CONSTANTS (k)		
	DRILLING	MILLING	TURNING
MILD STL. 25 RC	1.0	1.0	.9
MILD STL. 25 - 30 RC	1.6	1.8	1.3
HARD STL. 50 RC,	1.9	2.1	1.5
SOFT CAST IRON	.8	.7	.5
HARD CAST IRON	.9	1.1	1.0
ALUMINUM	.35	.4	.3
BRASS	.5	.6	.4
BRONZE	.6	.8	.7
STAINLESS 400	1.3	1.3	1.1
STAINLESS 300	1.6	1.8	1.7
TITANIUM	1.0	1.0	1.1
NICKEL ALLOYS	1.6	1.6	1.5

Figure 4.2   Table of constants.

4.1.3 Horsepower At Spindle

$$HPs = k \cdot Qd \text{ at } 100\% \text{ efficiency}$$

Where

$$HPs = \text{Horsepower required for machining}$$

$$Qd = \text{Volume of material removed in in}^3/\text{min or cm}^3/\text{min}$$

$$k = \text{Material constant, as shown in Figure 4.2.}$$

Since 100% efficiency is a theoretical value, we must in fact alter the above equation in view of the efficiency (E) of the spindle drive. In practice, E may vary from $0.7 \sim 0.85$ depending on the condition of the machine tool. The practical formula for calculating the horsepower (HP) will therefore change to:

$$HPm = \frac{k \cdot Qd}{E}$$

where HPm represents the horsepower of the spindle drive motor.

4.1.4 Torque on Spindle Due to Drilling (Ts)

$$Ts = \frac{63030 \cdot HPs}{n}$$

(NOTE: Horsepower is the unit of power adopted for engineering use. HP = 33.000 ft lb/min = 550 ft lb/sec. Metric horsepower = 75 kg · m/sec = 542.5 ft lb/sec. While the International System (metric) unit of power is the watt, its applications haven't yet reached the usual engineering tables.)

4.1.5 Machining Time

Machining time is expressed in terms of minutes.

$$T = \frac{L}{F}$$

where

$$L = \text{Depth of drilling in inches or mm}$$
$$F = \text{Feed in ipm or mm/min}$$

4.2 TURNING

4.2.1 Cutting Speed (Vt)

See Figure 4.3 for illustration of turning.

$$Vt = \frac{D \cdot \pi \cdot n}{12} \text{ fpm in inch units}$$

or

$$Vt = \frac{D \cdot \pi \cdot n}{1000} \text{ m/min in metric}$$

4.2.2 Rate of Metal Removal (Qt)

$$Qt = T \cdot F \cdot n \cdot c \qquad \text{in in}^3/\text{min or cm}^3/\text{min}$$

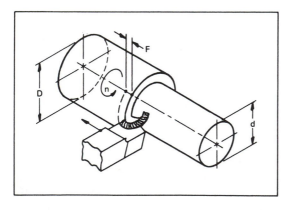

Figure 4.3   Turning.

Where

$$Qt = \text{Volume of material removed}$$

$$F = \text{Feed in ipr or mm/rev}$$

$$n = \text{Part rpm}$$

$$T = \text{Depth of cut } \frac{D - d}{2}$$

$$C = \text{Circumference } D \cdot \pi$$

4.2.3 Horsepower

$$HPm = \frac{k \cdot Q}{E}$$

HPm is the horsepower of the spindle drive motor.

4.2.4 Torque on Headstock due to Turning

$$Ts = \frac{63030 \cdot HPs}{n}$$

4.2.5 Surface Roughness

The theoretical surface roughness (Sr) can be calculated by the formula below. However, this should only be used as a programming guideline. Actual surface finish depends on a number of other factors, such as sharpness of tool, coolant, tool part rigidity, etc. Therefore, the calculated result may vary substantially if any or all of the above factors are not ideal.

$$Sr = \frac{F}{8000 \cdot r} \text{ microinches, or}$$

$$Sr = \frac{F}{8 \cdot r} \text{ micrometres}$$

Where

$$F = \text{Feed rate in ipr or mm/rev}$$

$$r = \text{Tool nose radius in mm or inches}$$

4.2.6 Acceleration and Deceleration Distance for Thread Turning

The programmer must allow a distance A_1 for the tool to accelerate to a constant (programmed) feed rate and a distance A_2 for deceleration. See Figure 4.4. Since the tool point velocity (feed rate) will change from zero to the programmed rate in the A_1 interval, the thread cut on this distance will be imperfect. The same condition applies to the A_2 distance.

The formulas below can be used as a guideline if specific formulas are not provided by the manufacturer of your CNC system.

The value of the time constant t_1 of the system, provided by the manufacturer, is a function of the specific system dynamics. Factors such as the weight of the tool slide or turret, coefficient of friction, size and torque of the servodrive, servogain, etc., influence the specific CNC time constant.

The Fanuc 4NE system, used in our example, has a time constant $t_1 = 0.12$. Multiplying this by the linear velocity of the tool point, $A_2 = t_1 \cdot V_L$, allows us to calculate the distance required to decelerate the tool from the programmed feed (velocity) to zero.

Example

Calculate the deceleration distance required to turn a 1½″–8 tpi thread at 350 rpm.

Solution

$$8 \text{ tpi} = 0.125 \text{ ipr},$$

The linear velocity of the tool point will be 0.125 inch for each revolution.
The linear velocity of the tool point at 350 rpm:

$$V_L = 0.125 \cdot 350 = 43.75 \text{ ipm} = 0.729 \text{ ips}$$

$$A_2 = t_1 \cdot V_L = 0.12 \cdot 0.729 = 0.0875 \text{ inches}$$

which is the deceleration distance.

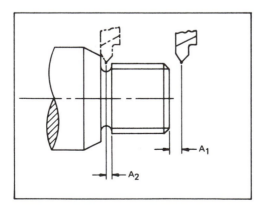

Figure 4.4 Tool point acceleration-deceleration.

The acceleration distance calculation requires a value for the acceleration time constant, t_2. This is calculated as follows:

$$t_2 = t - t_1 + t_1 \cdot EXP \left(\frac{-t}{t_1} \right)$$

Where

$$EXP \left(\frac{-t}{t_1} \right) = 0.05$$

EXP is the abbreviated version of "e at the power of . . . ," where $e = 2 \cdot 7182818$.

Using natural logarithms on both sides of the equation, we obtain

$$\frac{-t}{t_1} = \ln 0.05 = -2.99573$$

$$-t = t_1 \cdot (-2.99573)$$

or

$$t = 0.12 \cdot 2.99573 = 0.3595$$

Substituting this value into the above equation we obtain:

$$t_2 = 0.3595 - 0.12 + 0.3594 \cdot 0.05 = 0.2455$$

The acceleration distance can now be calculated from the formula:

$$A_1 = t_2 \cdot V_L$$

Example

Calculate the acceleration distance required to turn a 1½"–8 tpi threads at 350 rpm.

Solution   As in the preceding example.

$$8 \text{ tpi} = 0.125 \text{ ipr}$$

$$V_L = 0.125 \cdot 350 = 43.75 \text{ ipm} = 0.729 \text{ ips}$$

and

$$A_1 = t_2 \cdot V_L = 0.2455 \cdot 0.729 = 0.1789 \text{ inches}$$

This is the acceleration distance.

The A_1 distance varies between 2 to 4.4 times the A_2 for CNC systems, or

$$A_1 = 2 \sim 4.4 \cdot A_2$$

4.3 MILLING

4.3.1 Cutting Speed (Vm)

$$Vm = \frac{D \cdot \pi \cdot n}{12} \text{ fpm, where the diameter } D \text{ is expressed in inches.}$$

and

$$Vm = \frac{D \cdot \pi \cdot n}{1000} \text{ m/min}$$

where D is in mm.

4.3.2 Rate of Metal Removal (Qm)

$$Q_m = W \cdot T \cdot F \text{ in}^3/\text{min or}$$

$$Q_m = \frac{W \cdot T \cdot F}{1000} \text{ cm}^3/\text{min(metric)}$$

where

Q_m = Volume of metal removed in in^3/min or cm^3/min, as applicable.

W = Width of cut in inches or mm

T = Depth of cut in inches or mm

F = Programmed feed in ipm or mm/min.

4.3.3 Horsepower

$$HPm = \frac{k \cdot Q}{E}$$

4.3.4 Torque on Spindle

Torque on the spindle can be calculated as

$$Ts = \frac{63030 \cdot \text{HPs}}{n}$$

The calculations and formulas presented in this chapter are for ideal machining conditions. The reader must adjust the calculated values if the rigidity of the setup is not as good as desired or if the tool deflection, due to the cutting forces, create undesirable (destructive) vibrations.

5

Cutter Centerline
Programming

Before getting involved in the actual programming, we shall briefly review some of the very basic codes and principles.

Absolute Programming. This is a mode of programming in which an origin has to be selected for each axis prior to starting the program. Once this "part program origin" has been selected, all motions have to be stated with respect to this origin. What this means is that all motions defined in the program are in reality "locations" or "addresses" of the particular point defined in relation to the origin. Usually this is the mode of programming selected by experienced programmers, as the programmed values match fairly closely the values defined on the engineering drawing. The drawing's datum point is a direct equivalent of the program origin.

In milling programming, the absolute mode is established by the use of the preparatory code G90. In turning, the use of the tape words X and Z means that the cross direction and the longitudinal direction are programmed in absolute, without requiring a particular preparatory code.

Incremental Programming. When programming in incremental, an origin need not necessarily have been selected. All motions are stated from the immediate last position of the tool. This means that all incremental motions are "displacements" from a given position. The immediate short-range advantage of incremental programming is that the programmed motion matches directly the actual motion of the tool, as both take place from the last previous position of the cutter. Another advantage is that the sign of the motion is also directly related to the tool motion. In incremental programming, a positive dimension will cause a tool motion in the appropriate direction, and a negative one will take place in the opposite sense. In absolute programming, the sign depended on the quadrant the tool moved in, not in the direction of travel.

In milling programming, incremental is programmed using the preparatory code G91. In turning, incremental may be programmed the same way, or using the letters U and W to replace X and Z, in which case the preparatory code G91 is not required.

Programming the Origin. Establishing the origin of the part program for subsequent use in absolute programming is known under several terms such as *register preset, work origin setting, program zero point,* or *position absolute coordinates setting.* This proliferation of terminologies exists because NC technology is fairly new, and yet it has expanded too rapidly to allow standardization to catch up with it.

A program origin and a coordinate system are a must. The program origin preparatory code, G92 in milling and G50 in turning, will not cause any motion. The code will tell the system where the tool is located at that given time in relation with the program zero. In reality, the machine knows where the tool is and does not know where the program origin is. Therefore, when we tell the machine where the tool is in relation with the work zero, we are really telling it where the work origin is in relation to the known position of the tool. From there on, absolute programming will give addresses from this datum point.

Rapid. The G00 (G zero zero) will result in rapid positioning of the programmed machine slides to the required location. It is used for rapid approaching of the part, or for rapid moves between holes in drilling or boring applications. The tool or the table will always travel at the highest machine speed. It should be borne in mind, however, that regardless of the actual element of the machine that provides the move, in our program it is the tool that moves in relation to the part.

In most controls (there are some recent exceptions), the rapid motion will follow the "point-to-point" pattern, i.e., all programmed axes will start simultaneously, and as one is achieved, its motor will stop while the others will continue as required.

A safety reminder may be in order for people who sometimes lean on machine components. It is not unusual for a CNC machine to move in Rapid at 10 ips (approximately 250 mm/s).

Linear Interpolation. This feature is programmed using the preparatory code G01 (G zero one), in conjunction with the appropriate dimensional tape words, as well as a programmed feed. The corresponding machine axes, with their own variable speed-controlled drive systems independent from each other, will produce the required straight-line motions by driving the slides at different speeds.

Circular Interpolation.   Two codes are used in programming a circular arc, G02 (clockwise) and G03 (counterclockwise). The CNC system has the capability to establish and maintain the relative positions and velocities of two machine slides, on a constantly changing basis, but starting and stopping at the same time.

At the start of circular interpolation, the control knows the position of the cutter. It must be told the desired position at the end of the programmed arc, in either incremental or absolute coordinates. The control also has to know the location of the center of arc and the value of the arc radius. This may be achieved using I and J or K tape words in older controls, or R (for radius) in the newer ones. Subsequent programming examples in this chapter will illustrate this particular technique.

This chapter will present, in some detail, cutter centerline programming. Its importance is underscored by the fact that older systems can only be programmed in cutter centerline. Newer systems, with advanced cutter compensation features, are much easier to comprehend if cutter centerline programming is known and understood. In addition, in some special cases, of very intricate parts, the use of the compensation features may present a complication, and the programmer can solve the problem and get the job done by using an "old-fashioned" programming method on a state-of-the-art CNC system.

The program will guide the cutter around the part contour. The cutter will have to follow the path at a set distance away from the part, at every point, corresponding to the cutter radius.

Parts with complex geometry will require a certain level of calculations, mainly trigonometry, so a review of the math chapter may be necessary.

5.1 CALCULATING CUTTER CENTERLINE DISTANCES

To write a part program for the outside contour milling of the job illustrated in Figure 5.1, we shall first have to select a cutter (end mill) diameter.

In cutter centerline mode, the cutter center is programmed, yet the actual cutting is performed by the edge of the cutter. In order to obtain the required shape of the part, the cutter must be correctly placed at each point (1, 2, 3, . . . to 8 and back to 9) and of course in between, as shown in Figure 5.1.

This placement is that of a circle (i.e., the cutter) tangent to a line or circle representing the part contour.

The line connecting the cutter center to the cutting edge at the part (cutter radius) will be perpendicular to the part contour being cut.

This requirement of perpendicularity at a known location, combined with the knowledge of the cutter radius (half the cutter diameter) will enable us to calculate the cutter center location at each point by solving appropriate right triangles, as shown later in the chapter. The program drives the machine spindle where the center coincides with the cutter center.

Bear in mind that regardless whether the cutting action is achieved by moving

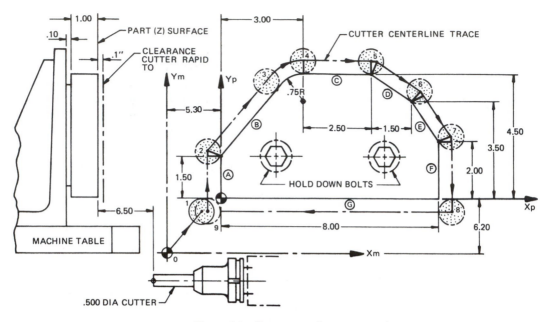

Figure 5.1   Cutter centerline programming.

the cutter or the part, in programming it is the cutter that is assumed to be mobile
and the part that is considered stationary. Always *think tool!*

Having calculated these dimensions, we must then convert them into programmable distances. The actual values of these distances will be different depending on whether we use incremental or absolute programming. The calculation
of the values may take up more time than the actual programming, and the reader
may be tempted to skip straight to cutter compensation or computer-assisted programming. However, neither can be properly understood and used without a strong
foundation in cutter centerline programming.

In the following, we shall discuss the necessary steps required to calculate
the center point locations of a 0.500-inch-diameter cutter, in "incremental" programming mode. We will also assume that the part has been bandsawed to size,
leaving approximately 0.150 inch for finish machining, and that the material is SAE
1030.

The programming methods illustrated below can be used for any similarly
shaped part so long as the machine codes have been double-checked for the actual
system used.

START UP
    N0010 G20 G40 G80 G91   Inch programming (G20 or G70),
                               cancel cutter diameter compensation,
                               cancel canned cycles, incremental
                               programming. The "cancels" are for
                               safety.

As mentioned at the beginning of this chapter, inexperienced programmers should start programming in incremental (G91). The incremental motion is a "displacement" of the tool from its present location, as opposed to absolute, which takes the tool to an address or a location defined in relation to an origin. Incremental provides a direct correlation between programmed motion and tool motion.

N0020 G92 X0 Y0 Z0	Set control display to zero, to correspond to the machine "zero" position.
N0030 S800 M03	Spindle speed 800, spindle start clockwise
N0040 G00 X4.75 Y5.95	"Rapid" to part in X and Y
N0050 Z-6.4	Rapid to part in Z, leaving 0.1 inch clearance
N0060 G01 Z-1.15 F3.0	Feed 0.05 inch past part at 3 ipm
N0070 X0.3 M08	Feed to point 1, coolant on

5.1.1 Machine Part Surface "A" Between Point 1 and Point 2

Since the necessary "Y" motion is not readily available on the part drawing, we shall have to calculate it from the given dimensions. There is no "X" motion from point 1 to point 2. However, the "Y" motion will be longer than the 1.5-inch length of surface "A" by an amount $\Delta Y2$ (see Figure 5.2).

This calculation is necessary in order to position the cutter radius "r" perpendicular to the upcoming surface "B"

$$\Delta Y2 = r \cdot \tan \left(45° - \frac{\alpha 1}{2} \right) \tag{5-1}$$

$$\tan \alpha = \frac{2.25}{3.00} = 0.75$$

$$\alpha = 36.8699°$$

$$R = \sqrt{3.0^2 + 2.25^2} = 3.750$$

$$\sin \alpha 2 = \frac{0.75}{3.75} = 0.2$$

$$\alpha 2 = 11.5369°$$

$$\alpha 1 = \alpha + \alpha 2 = 36.8699 + 11.5369 = 48.4068°$$

Substituting $\alpha 1$ into equation (5-1), we obtain

$$\Delta Y2 = r \cdot \tan \left(45 - \frac{48.4068}{2} \right)$$

$$= 0.25 \cdot \tan 20.7966 = 0.0949 \text{ inch}$$

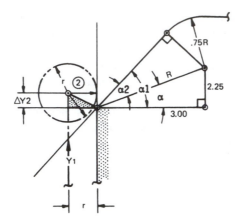

Figure 5.2   Cutter center at point 2.

Having calculated the $\Delta Y2$ value, we can now derive the $Y2$ value, to be programmed as the "Y" motion from point 1 to point 2.

$$Y2 = r + 1.5 + \Delta Y2 = 0.25 + 1.5 + 0.0949 = 1.8449 \text{ inches}$$

and the next program line will be

> N0080 Y1.8449   Move cutter, in feed, to point 2

5.1.2 Machine Part Surface "B" to Point 3 and Arc to Point 4

We must now calculate the dimensions $X3$ and $Y3$ from Figure 5.3, prior to programming the tool motion from point 2 to point 3. These dimensions again cannot be readily taken off the drawing, as is the case in most programming applications. Calculations similar to the ones above are required for arc to straight-line translations.

From the previous calculations we retain:

$$\alpha1 = 48.4068°$$

$$\Delta Y2 = 0.0949 \text{ inch}$$

$$r = 0.25 \text{ inch (the cutter radius)}$$

$$R = 0.75 \text{ inch (the part radius)}$$

$$\Delta Y3 = r \cdot \cos \alpha1 = 0.25 \cdot \cos 48.4068° = 0.1659 \text{ inch}$$

$$\Delta J = R \cdot \cos \alpha1 = 0.75 \cdot \cos 48.4068 = 0.4978 \text{ inch}$$

$$Y3 = 2.25 + \Delta J + \Delta Y3 - \Delta Y2$$

$$= 2.25 + 0.4978 + 0.1659 - 0.0949 = 2.8188 \text{ inches}$$

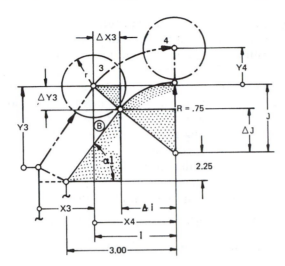

Figure 5.3 Cutter centerline geometry at straight-line and circular arc intersection.

This will be the Y-motion from point 2 to point 3.

$$\Delta Y = r \cdot \cos \alpha 1 \qquad\qquad (5\text{-}2)$$

$$\Delta X = r \cdot \sin \alpha 1 \qquad\qquad (5\text{-}3)$$

$$\Delta j = R \cdot \cos \alpha 1 \qquad\qquad (5\text{-}4)$$

and

$$\Delta i = R \cdot \sin \alpha 1 \qquad\qquad (5\text{-}5)$$

$$j = \Delta j + \Delta Y3$$

and

$$i = \Delta i + \Delta X3$$

$$\Delta X3 = r \cdot \sin \alpha 1 = 0.25 \cdot \sin 48.6068° = 0.1875 \text{ inch}$$

$$\Delta j = R \cdot \sin \alpha 1 = 0.75 \cdot \sin 48.6068° = 0.5625 \text{ inch}$$

$$X3 = 3.0 + r - (\Delta X3 \cdot \Delta i)$$

$$= 3.0 + 0.25 - (0.1875 + 0.5626)$$

$$= 3.25 - 0.7501 = 2.4999 \text{ inches}$$

This will be the X-displacement from point 2 to point 3, and the next program will be

N0090 X2.4999 Y 2.8188   To point 3.

To machine the 0.750-inch radius arc, we have to program the X- and Y-displacements from the start of the arc to the end point of the arc, as well as the values "i" and "j," required to position the center of arc in relation to the start point of the arc. Point 3 is not on either of the primary axes (X or Y) of the system of coordinates. It may be worth pointing out at this time that as NC systems changed, they became progressively easier to program. Unlike many of the old machines still in use, the modern CNCs have no constraints of primary axes or i and j unit vectors for circular interpolation, trailing zeros, etc. Many of the newer systems accept the "old" formats as well, but the contrary is not true. Circular interpolation is normally programmed using direct radius programming, which is much easier. This program, however, will be done the hard way.

The calculations for the next tool motion are illustrated in Figure 5.3.

$$X4 = \Delta i + \Delta X3 = 0.5626 + 0.1875 = 0.7501 \text{ inch}$$

This will be the X-displacement from point 3 to point 4.

$$Y4 = R + r - (\Delta j + \Delta Y3) = 0.75 + 0.25 - (0.4978 + 0.1659)$$

$$Y4 = 0.3363 \text{ inch}$$

This will be the Y-displacement from point 3 to point 4

$$i = X4 = \Delta i + \Delta X3 = 0.7501 \text{ inch}$$

The dimension "i" is called a "unit vector," and it is measured from the start point 3 of the arc, where the cutter is located, to the center of the arc, to be measured parallel to the X-axis.

$$j = \Delta j + \Delta Y3 = 0.4978 + 0.1659 = 0.6637 \text{ inch}$$

The dimension "j," also called a "unit vector," is measured parallel to the Y-axis from the start point of the cutter to the center of the arc. For the reader who has not at this point developed an understanding of the circular interpolation process, the following explanation may be helpful. The cutter is in position prior to the start of the arc, and the control knows its position, since this was the end point of the previous travel motion. From the start point of the arc, the control will travel an imaginary path along "j," then along "i." At this point the computer has found the center of the arc. Using the values i and j as sides of a right-angle triangle, the control will compute the hypotenuse of the triangle, which is the radius of the arc. Knowing the center and the radius, the control will now trace an imaginary circular path, along which the cutter will travel a real distance defined by the programmed dimensions X and Y.

The circular interpolation motion block can be written therefore as follows:

N0100 G02 X0.7501 Y0.3363 I0.7501 J-0.6637

The unit vectors can be spelled in either lowercase or capital letters.

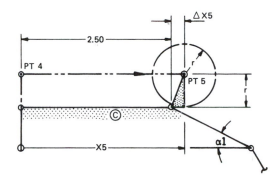

Figure 5.4 Cutter centerline geometry at point 5.

5.1.3 Machine Part Surface "C" from Point 4 to Point 5

The calculations for machining the part surface "*C*" from point 4 to point 5 will be identical to the ones corresponding to point 2.

The offset will be in the *X*-direction by the amount $\Delta X5$.

$$\Delta Y5 = r$$

$$\Delta X5 = r \tan \frac{\alpha 1}{2} \tag{5-6}$$

From the part dimensions,

$$\text{Tan } \alpha 1 = \frac{1.00}{1.50} = 0.66666, \text{ or } \alpha 1 = 33.59°$$

Using the triangle shown in Figure 5.4 and formula (5-6),

$$\Delta X5 = r \cdot \tan \frac{\alpha 1}{2} = 0.25 \cdot \tan \frac{33.59}{2} = 0.0754 \text{ inch}$$

$$X5 = 2.50 + \Delta X5 = 2.50 + 0.0754 = 2.5754 \text{ inch}$$

This will be the *X*-motion from point 4 to point 5, and the corresponding program line will read:

N0110 G01 X2.5754

5.1.4 Machine Surface "D" from Point 5 to Point 6

To machine the surface "*D*" from point 5 to point 6, the cutter motion will require offsets in both *X*- and *Y*-directions, since neither surface "*D*" nor "*E*" is parallel with the primary axes of the system. The angles $\alpha 1$ and β will have to be measured

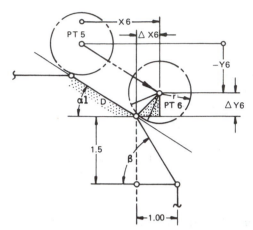

Figure 5.5 Cutter centerline geometry at point 6.

from a theoretical line, parallel with the X-axis, as shown in Figure 5.5. From Figure 5.4, $\alpha1 = 33.59°$

$$\Delta Y6 = \frac{\cos \dfrac{\alpha1 + \beta}{2}}{\cos \dfrac{\alpha1 - \beta}{2}} \cdot r \tag{5-7}$$

$$\Delta X6 = \frac{\sin \dfrac{\alpha1 + \beta}{2}}{\cos \dfrac{\alpha1 - \beta}{2}} \cdot r \tag{5-8}$$

$$\text{Tan } \beta = \frac{1.5}{1} = 1.5 \quad \text{or} \quad \beta = 56.3099°$$

$$\Delta Y6 = \frac{\cos \dfrac{\alpha1 + \beta}{2}}{\cos \dfrac{\alpha1 - \beta}{2}} \cdot r = \frac{\cos \dfrac{33.59 + 56.31}{2}}{\cos \dfrac{33.59 - 56.31}{2}} \cdot 0.25$$

$$= \frac{\cos 44.95}{\cos (-11.36)} \cdot 0.25 = \frac{0.707724}{0.980409} \cdot 0.25 = 0.1804 \text{ inch}$$

$$\Delta X6 = \frac{\sin \dfrac{\alpha1 + \beta}{2}}{\cos \dfrac{\alpha1 - \beta}{2}} \cdot r = \frac{\sin 44.95°}{\cos (-11.36)} \cdot r = \frac{0.706489}{0.980409} \cdot 0.25$$

$$= 0.1801 \text{ inch}$$

$$X6 = 1.5 + \Delta X6 - \Delta X5 = 1.5 + 0.1801 - 0.0754 = 1.6047 \text{ inch}$$

and

$$Y6 = 1.00 + r - Y6 = 1.00 + 0.25 - 0.1804 = 1.0696 \text{ inch}$$

The next tape block will be:

N0120 X1.6047 Y-1.0696

5.1.5 Machine Surface "E" to Point 7

The offset at point 7 can be calculated using equation (5-1). As an exercise, it is suggested the reader draw an enlarged figure of the geometry at point 7, while following the calculations below:

$$\Delta Y7 = r \cdot \tan \left(45\,° - \frac{\alpha 1}{2} \right) \qquad \beta \text{ will replace } \alpha 1$$

$$\Delta Y7 = 0.25 \cdot \tan \left(45° - \frac{56.31}{2} \right) = 0.25 \cdot \tan 16.845° = 0.0757 \text{ inch}$$

$$X7 = 1.00 + r - \Delta X6 = 1.00 + 0.25 - 0.1801 = 1.0699 \text{ inches}$$

$$Y7 = 1.5 + \Delta Y6 - \Delta Y7 = 1.50 + 0.1804 - 0.0757 = 1.6047 \text{ inches}$$

The program line will read:

N0130 X1.0699 Y-1.6047

5.1.6 Machine Surface "F" from Point 7 to Point 8

$$X8 = 0$$
$$Y8 = 2.00 + r + \Delta Y7 = 2.00 + 0.25$$
$$+ 0.0757 = 2.3257 \text{ inches}$$

The program line is:

N0140 Y-2.3257

5.1.7 Machine Surface "G" from Point 8 to Point 9

$$Y9 = 0$$
$$X9 = 8.0 + r + 0.05 = 8.0 + 0.25 + 0.05 = 8.3 \text{ inches}$$

The part would be machine finished without the additional 0.05-inch-increment. It is, however, a desirable practice to move the cutter past the finished surface by 0.01 to 0.1 inch.

The line of program corresponding to surface "G" is

N0150 X-8.3

5.1.8 Return Tool to the Machine Zero ("Home") Position

N0160 Z1.15 F0.4 M09	Clear part, stop coolant
N0170 G28 Z0.1	Return Z "home"
N0180 G28 X-0.1 Y-0.1	Return X and Y "home"

The few basic calculations discussed so far will not cover more complex part surface intersections, such as the tangency of two circular arcs or an arc tangent to a line. These will come up later, when more complex part geometries will be studied under CNC turning.

5.2 TOOL NOSE RADIUS CENTERLINE CALCULATIONS FOR CNC TURNING

Tool nose radius (TNR) calculations are also known as "equidistance programming." When CNC turning is first discussed, TNR is usually overlooked. However, once the basics have been established, we must remember that every turning tool tip has a radius, small as it may be. See Figure 5.6. This radius will have to be taken into consideration when the turned part surfaces are not parallel with the primary axes of the machine (X and Z in the case of the turning center). The error will be the greatest when turning a surface at 45° from the primary axes.

The actual inaccuracy or error, E, can easily be calculated if we know the radius r of the tool tip. Assuming a value of $r = 0.015$ inch, for example, the error

$$E = H - r = \frac{r}{\sin 45°} - r = \frac{0.015}{0.7071} - 0.015 = 0.0062 \text{ inch}$$

The value of

$$H = \frac{r}{\sin 45°}$$

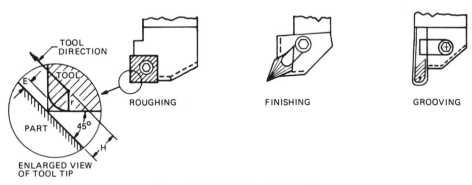

ROUGHING   FINISHING   GROOVING

Figure 5.6   Typical turning tool tips.

results from Figure 5.6. The tapered part surface diameter at a 45° angle would therefore be larger by 2 · 0.0062 = 0.0124 inch. This deviation would not be acceptable from a quality control point of view.

It can be seen at this stage that this inaccuracy problem is more complex in the case of turning than it is for milling, because of its variability. The error disappears completely when turning surfaces parallel to either the X- or Z-axis of the turning center. For tapered surfaces, the error becomes proportional to the part surface angle, the deviation increasing from zero part surface angle to its maximum value at 45°. In addition, the error will further increase as the TNR becomes larger.

To eliminate this tool tip error in CNC turning systems without programmable TNR compensation, we must, prior to programming, calculate the path of the tool tip radius center. The method of calculation is similar—in fact in many cases identical—to the one discussed in section 5.1.

To illustrate the different steps to follow, the equidistance calculations will be carried out in parallel with the programming for the part shown in Figure 5.7.

In the following, we have used the *right-hand coordinate system*, common to most CNC turning centers. In the right-hand system, $+Z$ points away from the chuck, $+X$ points away from the operator, G02 is circular interpolation clockwise, G03 is circular interpolation counterclockwise, G41 is tool nose radius compensation left and G42 is tool nose radius compensation right. The standard tool nose vector chart is as illustrated in Figure 7.6.

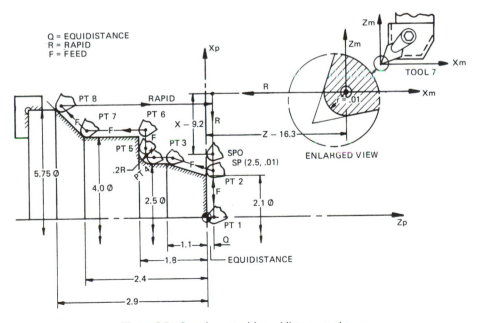

Figure 5.7 Sample part with equidistance tool trace.

Some turning centers use the *left-hand coordinate system*. The essential difference is that $+X$ now points towards the operator. This change reverses G02 which becomes counterclockwise, G03 which becomes clockwise, G41 which becomes right, and G42 which becomes left. The changes in the chart from Figure 7.6 are as follows: point 2 changes place with 3, 1, with 4, and 6 with 8.

As in the case of milling, it will be assumed that the part has been previously machined, in our case rough-turned. The calculations will be performed and the program written for the turning of the finished contour as dimensioned. Equidistance calculations are required from point 1 through to point 8. As we proceed with the calculations and the program, the geometry will be drawn in sufficiently enlarged scale so that the necessary trigonometric labeling can be clearly analyzed for each step along the path.

5.2.1 Start-Up

It will be assumed that the tool slides have been brought to "home" position.

N01 G50 X0 Z0	"Zero" control to machine home
N05 M41	Select middle speed range
N10 G98 G20 T0700	Feed in, ipm inch programming, turret tool position no. 7
N15 S950 M03	Spindle speed 950 rpm, turn spindle on clockwise.
N20 G00 Z-16.29	Rapid to point 0 in Z
N25 X-9.2	Rapid to point 0 in X (diameter programming)

Having moved the tool into position for cutting in the machine coordinate

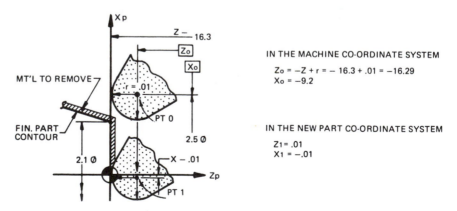

IN THE MACHINE CO-ORDINATE SYSTEM

$Z_o = -Z + r = -16.3 + .01 = -16.29$
$X_o = -9.2$

IN THE NEW PART CO-ORDINATE SYSTEM

$Z_1 = .01$
$X_1 = -.01$

Figure 5.8   Equidistance at points 0 and 1.

system, we should, at this point, transfer the coordinate system onto the part as shown in Figure 5.8. This will allow us to program all the necessary slide motions in terms of the actual part dimensions. The calculated equidistances will thus be added to or subtracted from actual part dimensions, rather than worked out in terms of dimensions from the machine "home" position.

> N30 G50 X2.5 Z0.01   Established new coordinates in terms of
> the part coordinate system XpZp. There
> was no tool motion involved.
>
> N35 G01 X-0.01 F2.0   Turn face

5.2.2 Move Tool to Point 2

The next surface being tapered, the X-position of the tool must be calculated. As shown in Figure 5.9, no calculations are required for Z, whose position remains tangent to the part face.

The following important geometric condition arises at point 2: The TNR must be tangent to both current and upcoming part surfaces. From equation (5-1),

$$\Delta X2 = r \cdot \tan \left(45° - \frac{\alpha}{2} \right) = 0.01 \cdot \tan \left(45° - \frac{10.305}{2} \right) = 0.0083 \text{ inch}$$

where α was calculated as $\alpha = \text{InvTan} \dfrac{(2.5\text{-}2.1)/2}{1.1} = \text{InvTan } 0.18181 = 10.305°$

and the next block of tape will be:

> N40 X 2.1166 F15.0   Move to point 2 (the value of the X-
> coordinate was calculated in Fig. 5-9)

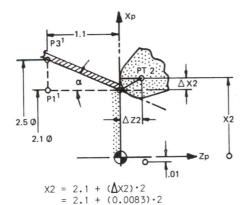

X2 = 2.1 + (ΔX2)·2
   = 2.1 + (0.0083)·2

X2 = 2.1166

Figure 5.9 Equidistance at point 2.

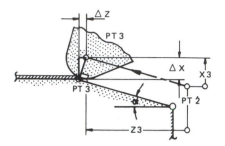

Figure 5.10 Equidistance at point 3.

(NOTE: Your calculator may be labeled Invtan, Arctan, Tan^{-1}, or it may use a function key to calculate an angle when the trigonometric function is known.)

5.2.3 Turn Tapered Surface from Point 2 to Point 3

To maintain the equidistance condition at point 3, the $Z3$ motion will have to be adjusted by a calculated $\Delta Z3$.

As the surface between points 3 and 4 is parallel to the Z-axis, the X-adjustment will be made by adding the value of the radius to the corresponding dimension. Since X is programmed as a diameter, the adjustment will have to be $2r$.

The geometric conditions at point 3 are as follows:

- The tool tip radius r is perpendicular to the line between points 2 and 3.
- The tool tip radius is tangent at both surfaces, current and next.
- The tool motion between points 2 and 3 is not parallel to either of the primary axes. See Figure 5.10.

$$\Delta Z = r \cdot \tan \frac{\alpha}{2} = r \cdot \tan \frac{10.305}{2} = 0.0009 \text{ inch}$$

$$X3 = 2.5 + 2r = 2.5 + 0.020 = 2.520 \text{ inches}$$

This is the X-motion to point 3.

$$Z3 = 1.1 - \Delta Z = 1.1 - 0.0009 = 1.0991 \text{ inches}$$

This is the Z-motion to point 3. The next program block will therefore read:

N45 X2.52 Z-1.0991 F7.5

The inquisitive reader has noticed by now that the X- and Z-dimensions have been programmed all along in absolute mode. The calculated values have been added to or subtracted from the actual part dimensions which are defined from two datum lines (axes) intersecting at the part origin (see Figure 5.7).

5.2.4 Turn 2.5-Inch Diameter to Point 4 and 0.2-Inch Radius to Point 5

$X4 = X3 = 2.52$ inches; there are no changes in the X-direction

$Z4 = 1.8 - R = 1.8 - 0.2 = 1.6$ inches

The next program block will read:

N50 X2.5 Z-1.6

The program is in absolute mode, and the same motion can be programmed as follows:

N50 Z-1,6   Note that usually "R" refers to the
$X5 = 2.5 + 2 \cdot R = 2.5 + 0.4$;   part while "r" is the TNR
$X5 = 2.9$ inches
$Z5 = 1.8 - r = 1.8 - 0.01 = 1.79$ inches
$R = 0.2$

The next block reflects the ability of an advanced control to perform circular interpolation (clockwise) in "radius" programming. We could have, of course, used I and J unit vectors, as in the milling example, with the same practical results. However, the additional calculations may increase the chances of errors, and the additional "characters" and "words" will lengthen the program tape. See Figure 5.11.

N55 G02 X2.9 Z-1.79 R0.2

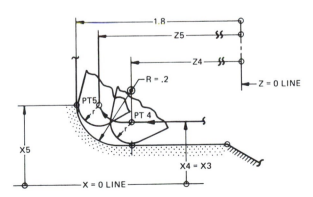

Figure 5.11 Equidistance at points 4 and 5.

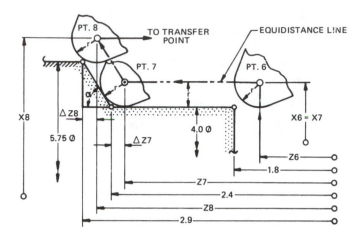

Figure 5.12   Equidistance at points 6, 7, and 8.

5.2.5 Turn Face to Dimension 1.8 Inches and 4.0-Inch Diameter to Point 6, Diameter to Point 7, and Tapered Surface to Point 8

By now, calculating these points should not present any difficulty. The intersecting points of the tool tip radius can easily be projected from the part surface lines. The geometry of the part-tool relationship is illustrated in Figure 5.12.

$$X6 = 4.0 + 2r = 4.0 + 0.02 = 4.02 \text{ inches}$$

$$Z6 = 1.8 - r = 1.8 - 0.01 = 1.79 \text{ inches}$$

The next block contains a G01 in order to return to linear interpolation from the preceding circular interpolation. Overlooking this G01 is one of the most frequently made mistakes.

The program block will look as follows:

N60 G01 X4.02 Z-1.79

There is no "Z-address" change, i.e., no Z-motion is required from point 5 to point 6. In absolute programming, the axis that does not contain a displacement can be left out, thus shortening the program. Accordingly, block 60 can be rewritten as shown below:

N60 G01 X4.02

Continuing the calculations,

$$\text{Tan } \alpha = \frac{\dfrac{5.75 - 4.0}{2}}{2.9 - 2.4} = \frac{0.875}{0.5} = 1.75, \text{ or } \alpha = 60.255°$$

$$\Delta Z7 = r \cdot \tan \frac{\alpha}{2} = 0.01 \cdot \tan 30.1275° = 0.0058 \text{ inch}$$

Accordingly,

$$Z7 = 2.4 - \Delta Z7 = 2.4 - 0.0058 = 2.3942 \text{ inches}$$

$X7 = X6$, and the motion to point 7 can be programmed as follows:

```
N65 Z-2.3942
```

The next block will be a linear interpolation in both axes. Therefore, the location of point 8 will also have to be calculated. If you observe the tool positions at points 7 and 8, you will note that the angle "α" is shared.

$$\Delta Z8 = \Delta Z7, \text{ already calculated to be } 0.0058 \text{ inch}$$

Hence $\Delta Z8 = 0.0058$ inch

$$X8 = 5.75 + 2r = 5.75 + 0.02 = 5.77 \text{ inch}$$

and finally

$$Z8 = 2.9 - \Delta Z8 = 2.9 - 0.0058 = 2.8942 \text{ inches}$$

The tape block for the motion to point 8 will read:

```
N70 X5.77 Z-2.8942
```

To conclude the program, the motions from the part system will be retransferred to the machine coordinate system as follows:

```
N75 G00 Z0.01              Motion to Z-transfer point
N80 X2.5 M09               Motion to X-transfer point, coolant off
N85 G50 X-9.2 Z-16.29      No motion, just transfer of the datum point.
N90 G28 X0 Z0 M05          Return "home," spindle off
N95 M30                    End of program
```

EXERCISES

5.1. Name some of the types of CNC machines that could be part of a flexible manufacturing system.

5.2. Estimate the degree of accuracy that could be produced by a contemporary CNC metal-cutting machine.

5.3. What are the two major assets of CNC machines?

5.4. Project. Draw a list of major items to be checked prior to acquiring a CNC machine. Prepare a table of these items for three comparable makes and models. Justify the selection for purchasing.

5.5. Define "absolute programming."

5.6. Define "incremental programming."

5.7. Explain "rapid" motion.

5.8. **(a)** What is the difference between linear and circular interpolation?
     (b) Explain each term in the following line of program:

N025 G01 X5.5 Y7.5

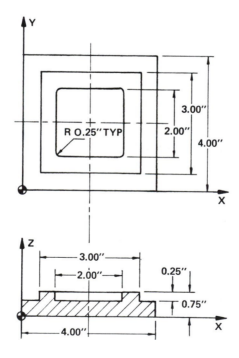

Figure 5.13 Milling sample.

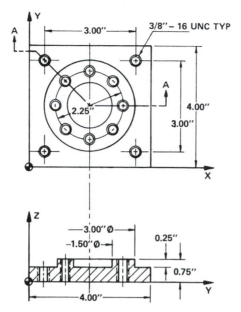

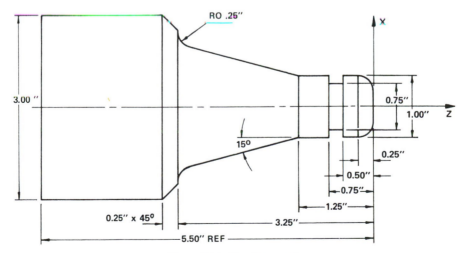

Figure 5.14 Milling, drilling, and tapping sample.

(c) Explain each term in the following line of program:

N035 G02 X4.00 Y5.25 I1.5 J2.5

(d) Calculate the radius of arc in sequence N035 (question 8c).

5.9. List some "nondimension" words in a program.

Figure 5.15 Turning sample.

5.10. List some "dimension" words in a program.

5.11. State a safety procedure for returning the tool "home" at the end of a CNC milling program.

5.12. What is a dry run? Give a few examples.

5.13. What is meant by M.D.I.?

5.14. What is the major advantage and the major drawback of cutter centerline programming, and what are some solutions to the problem?

5.15. Prepare setup and write cutter centerline programs for the parts in Figures 5.13, 5.14, and 5.15.

6

Tool Offsets

Most CNC machines have a set location, sometimes two, known as the "home" position, the "machine origin," or the "machine absolute origin." These expressions are used interchangeably. For the purpose of this book, we shall use the term "machine origin." The location is physically established by the manufacturer usually by a dual-limit switch system. The first one sets the motion deceleration, and the second one stops it.

The machine origin is used by manufacturers to synchronize the machine with the control, and to establish a start point for measuring the length of travel in the various axes. Some recent systems cannot even be started at all until this synchronization, known as the "zeroing," or "referencing," has been performed. The zeroing is carried out by returning the machine to "machine origin" and initializing the control manually or through programmed statements, such as G92 X0 Y0 Z0, or others, as applicable. By machine we understand the table, column, saddle, gantry, tool post, or whatever moving parts are involved. Initializing the control means setting the X, Y, Z, A, B, etc. displays (counters) to zero.

In addition to the machine origin, we have the following additional potential origins:

- the fixture origin
- the part origin
- the program origin

All these origins could be the same, or different, or grouped in some way. To show the flexibility of the systems, we could have more than one fixture or more than one part on the machine. In the latter case, the parts could be the same or different.

Over the years, as NC machines went through different phases, the term "zero offset" was used as an improvement over the "fixed zero," and sometimes called "zero shift," "full zero shift," or "floating zero." The objective was, of course, the ability to position the part at some advantageous location on the machine table and "offset" the "zero" to the right place.

The word "offset" has also been used in offset banks or offset registers. The terms "bank" and "register" are interchangeable, and they mean storage areas in the electronic control. These registers can be used for a number of purposes, such as producing small numerical changes to compensate for minute setup adjustments, dimensional inaccuracies, or tool wear. But they can also be used to store specific corrections related to a particular tool, tool radius, tool length, or one or more values related to a particular setup situation.

In some manuals or texts, cutter diameter compensation and cutter length compensation are quite often called "tool offset," "cutter offset," "tool length offset," etc.

In this chapter we will limit ourselves to the following uses of tool offsets:

- Tool offsets used for length compensation
- Flexible positioning of holding fixtures or parts
- Multiple part machining (same or different)
- Limited use in diameter compensation

These features are common to a large number of types of machines. The specific offsetting related to the ability to program the part rather than the tool will be called:

- Cutter diameter compensation (including the TNR compensation)
- Cutter length compensation

In these specific cases the term "compensation" will be used consistently in lieu of "offset," and detailed discussion of these important features will take place in Chapters 7 and 8.

6.1 TOOL OFFSET CODES USED FOR TOOL LENGTH COMPENSATION

Tool offset codes for tool length compensation are a very useful CNC programming feature that allow the operator to perform milling, drilling, tapping, or boring without presetting the tools to a specific length. As the name implies, this compensation controls the tool or "Z-" axis of a CNC machining center. The length of the compensation is controlled by the value stored by the operator in a specific offset register programmed for a particular tool.

It is a good technique to program a separate register for each tool length used. If the same tool is to be used in a different portion of the program requiring a different length, another register should be used. The basic principle is similar, in many cases identical, for most controls now being built.

Tool offsets used in tool length compensation essentially represent *addition* or *subtraction* by the CNC control. In the process of drilling, we program a specific rapid "Z-" motion to approach the part surface. In addition, we program a tool register number, such as H12, and a tool offset code G45 for lengthening (adding).

N0040 G00 G45 Z-3.0 H12

The interpretation of this tape block is the following:

- N0040—Sequence Number
- G00—Rapid Motion
- G45—Addition (tool offset)
- Z-3.0—Programmed Length of Tool Motion
- H12—Assigned Tool Register

This example is illustrated in the lower half of Figure 6.1.

We shall assume that the operator entered 2.9 into H12. The drill point will move in rapid motion by the amount of programmed "Z" *plus* the distance contained in tool register H12. It should be remembered that the tool will move in the *direction* given by the sign in "Z". When calculating, we add to the unsigned amount from "Z" the signed amount from the offset register. The resulting motion will take place in the direction of the sign resulting by combining the sign from "Z" with the sign of the added amounts. For example:

Z-3.0 H12 contains 2.9;	$3 + 2.9 = 5.9$;	motion -5.9
Z-3.0 H12 contains -2.9;	$3 - 2.9 = 0.1$;	motion -0.1
Z-3.0 H12 contains 3.9;	$3 + 3.9 = 6.9$;	motion -6.9
Z-3.0 H12 contains -3.9;	$3 - 3.9 = 0.9$;	motion $+0.9$

The tool offset code G46 will shorten (subtract) from the programmed motion the amount entered in the appropriate offset register.

N0050 G00 G46 Z-6.75 H11

Illustrated in the upper part of Figure 6.1, this example can be interpreted as follows: The drill point will move in rapid motion by the programmed amount

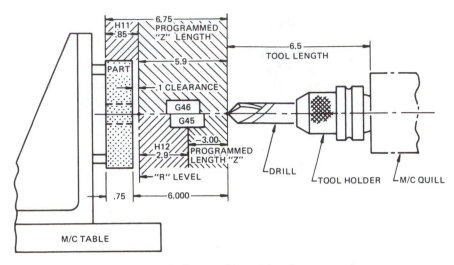

Figure 6.1 Tool offsets used in tool length compensation.

"Z" or 6.75 inches, *less* the distance 0.85 inch assumed entered in tool register H11.

In actual case, both programmed tape blocks in the examples above will result in the same tool point motion to "R" level, or 5.9 inches.

Even though the system accepts both tool lengthening or shortening codes, most programmers set a standard of using either one or the other, but not both. This will avoid potential confusion that could result in expensive tool crashes.

Using G45 in our example means that the operator can use any tool length less than or equal to 6.5 inches + 2.9 inches = 9.4 inches. At 9.4 inches of tool length, the value of H12 would be zero without having had to change the programmed Z-3.0 dimension.

Using G46, on the other hand, means that the operator can use any tool length less than or equal to 6.5 inches + 5.9 inches = 12.4 inches. At 7.5 inches of tool length, the value of H11 would be 1.85 inches, and at 10.5 inches of tool length the value of H11 would have to be changed to 4.85 inches. In other words, whatever we add to the tool length we have to add to the content of the tool register.

By now the reader may have concluded that it is easier to work with the G45 code. Let us assume that the shortest possible tool length (i.e., cutting tool and holder combination) is 6.5 inches. The clearance between tool and part at the end of the rapid approach is, for example, 0.1 inch. If we program a rapid motion of 1.0 inch, in conjunction with G45 and a particular tool offset register, the amount to be entered by the operator in that register will be the following:

The distance from the tool point to the part surface,

Less the programmed 1.0-inch motion,

Less the 0.1 inch-clearance.

6.2 TOOL OFFSETS USED FOR POSITIONING OF FIXTURE OR PART

Figure 6.2 illustrates a setup that requires the use of three different sizes of drills. Since we know how to write a part program using tool length compensation, we can ignore the actual length of our tools during the programming process. The following steps will outline the job:

1. The three drills specified are mounted in three different tool holders. The lengths are unimportant from a programming point of view, but rigidity, deflection, and proper grinding remain important and should not be overlooked.

2. The plate shown in Figure 6.2 is mounted on the machine table. The exact location is unimportant since we will use tool offsets in the X-Y plane. It is important to maintain the part edge marked "A" parallel to the table surface. The risers should be located so as to avoid tool interference.

3. The three tools are installed into the tool changer in the positions outlined by Figure 6.2.

4. The distances between the tool tip and the part surface are found as follows:
   - Position tool No. 1 in spindle.
   - Zero "Z" axis in "home" position above part.
   - Start spindle @ 100 rpm.

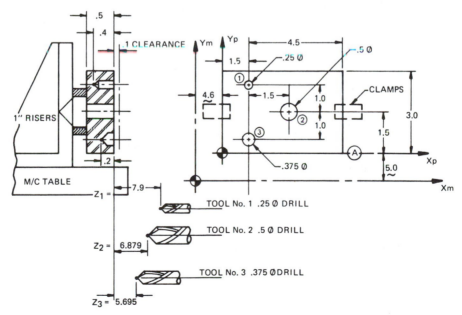

Figure 6.2 Sample part for tool offsets.

- Jog tool tip to touch the part.
- Read and note "Z" displacement. For our program, this reading is labeled $Z1$, and we will assume the respective amount to be 7.9 inches. The same exercise for tools Nos. 2 and 3 will yield $Z2 = 6.879$ and $Z3 = 5.695$.

5. The program for the drilling of the three holes shown in Figure 6.2 is now written below:

```
N0010 G20 G40 G80 G91      Initialization
N0020 G92 X0 Y0 Z0         "Zero" control
N0030 S1200 M03            Spindle on, 1200 rpm
N0040 G45 G00 X1.0 D01
N0050 G45 Y1.0 D02
```

Figure 6.2 shows two sets of coordinates. The origin of Xm and Ym is the machine origin, while the intersection of Xp and Yp is the part origin. From the machine origin, the coordinates of point "1" can be calculated from the drawing as $X = 6.1$ inches and $Y = 7.5$ inches. Had these dimensions been used in the program, the operator would have been required to set the part exactly to the dimensions $X4.6000$ and $Y5.0000$. As it stands, the tool offset registers D01 and D02 have been used. The operator has the choice of positioning the part any place he wishes, as long as the edge "A" is parallel to the table. Once the part is positioned, the operator measures the X- and Y-dimensions of the actual location. Since a displacement of 1.0 inch has already been programmed in sequences N0040 and N0050, the operator will subtract 1 inch from the X-amount and will enter the result in D01. He will subtract 1 inch from the Y-amount and will enter this result in D02. Due to G45, the amounts in the offset registers combined with the programmed motions will bring the tool to the required location.

If we assume that the part has been mounted at the operator's convenience and that the dimensions measured from the machine origin to the part origin are, respectively, 4.6 and 5.0 inches, as shown on the drawing, we have the following choices:

Along X,

$$4.6 + 1.5 = 6.1$$

Since we have programmed $X1.0$, we enter 5.1 in D01

Along Y,

$$5.0 + 1.5 + 1.0 = 7.5$$

We enter 6.5 in D02.

Alternatively, we could have programmed:

```
N0040 G45 G00 X1.5 D01
N0050 G45 Y2.5 D02
```

The operator measures 4.6 and 5.0 inches as before and enters them directly as measured in D01 and D02. The results are the same. We can now continue the program in the Z-axis.

        N0060 G00 G45 Z-1.0 H03   Tool length compensation

The value of the H03 offset register must at this point be calculated as: $Z1 - 1.0$ inch $- 0.1$ inch or 7.9 inches $- 1.0$ inch $- 0.1$ inch $= 6.8000$ where 1.0 inch is the Z-value programmed in sequence N0060 and 0.1 inch is the clearance.

N0070 G01 Z-0.5 F3.0 M08	Drill hole 1, 0.250-inch diameter, 0.4-inch deep
N0080 G00 Z0.5	Rapid return
N0090 G28 Z0	Return tool home
N0100 X1.5 Y-1.0 M05	Rapid to 2nd hole
N0110 M06 T02	Tool change
N0120 S900 M03	
N0130 G45 Z-1.0 H04	Tool length compensation

The value of the H04 offset register must be calculated exactly as in the case of H03 above.

N0140 G01 Z-0.63 F400	Drill 0.50 inch thru
N0150 G00 Z0.63	Rapid return
N0160 G28 Z0 M05	Return tool home
N0170 X-1.5 Y-1.0	Rapid to 3rd hole
N0180 M06 T03	Tool change
N0190 S1050 M03	
N0200 G45 Z-1.0 H05	Tool length compensation

The value of the H05 offset register must also be calculated as above.

        N0210 G01 Z-0.3 F350
        N0220 G00 Z0.3
        N0230 G28 Z0 M05
        N0240 G28 X 0 Y 0 M09
        N0250 M06
        N0260 M30

6.3 TOOL OFFSETS USED IN MULTIPLE PART MACHINING

This paragraph represents an extension of the preceding one. Let us assume that we're machining the part in Figure 6.2, but that in lieu of having positioned the part using tool offsets, we have positioned a holding fixture for that part. We are now machining a small production lot.

    In a jobbing shop, where production flexibility is vitally important, we have the ability to locate a second holding fixture in a different place on the table. The first fixture was located using D01 and D02. We can locate the second one using different offset registers, such as D11 and D12. We can now interrupt the first run, machine the second order, and then resume production. Where this type of requirement is a frequent occurrence, it is a good practice to program the *X*- and *Y*-motions first, with *Z* retracted all the way. This will avoid having to worry about crashing the tools for the second part while crossing the zone of the first one.

6.4 TOOL OFFSETS USED IN DIAMETER COMPENSATION

It is not recommended that offsets such as G45 or G46 be used in cutter diameter compensation. These codes will function in limited applications, for very simple geometries. Because errors may be introduced due to the design of the software, it is preferable to use G41 and G42, which have been specifically designed for this purpose. On specific controls, G41 and G42 may be used in conjunction with an additional code, G39. Cutter diameter compensation codes and their applications are discussed in subsequent chapters.

EXERCISES

6.1. Explain the term *origin* in programming context and list the different types of origins one can encounter in a program.

6.2. Explain the term *tool length compensation* and justify the need for this process.

6.3. What are the codes used in tool length compensation?

6.4. Given the following line of program, N30 G00 G45 Z-2.5 H10, calculate the tool motion if the values entered in register H10 are (a) 2.0 and (b) −2.0.

6.5. Is it important to know the actual length of the cutting tools before the start of the program?

6.6. Is it critical to position the fixture on the machine table exactly according to the setup drawing?

6.7. How can the compensated method of fixture positioning facilitate the flexibility of a small jobbing shop?

6.8. Write programs for the parts in Figure 5.13 and Figure 5.14 on pp. 90–91, using tool offsets for tool length compensation.

Cutter Diameter Compensation and TNR Compensation: Programming the Work Surface

7.1 CUTTER DIAMETER COMPENSATION

We have seen the complexities of cutter center programming in Chapter 5. While not difficult as such, programming the cutter center (sometimes called "programming the tool") can become involved, and the probability of errors increases with the number of calculations.

We shall now look at an entirely different approach. This consists of "programming the part." The programmer will write a part program tailored to the contour to be machined. This is equivalent to using an imaginary tool that has zero radius.

As part of the program, the programmer incorporates a "compensation" tape block. While the compensation "offsets" the tool, to minimize potential confusion in terminology, the process discussed in this chapter is "cutter diameter compensation" and not "tool offset." However, the numbered memory compartment where the value of the cutter radius will be stored is called "offset register." Offset registers are used in conjunction with different procedures such as tool offset, cutter diameter compensation, tool nose radius (TNR) compensation or tool length compensation.

The compensation tape block has the following functions:

1. To assign a specific offset register to a specific tool. CNC systems are normally equipped with programmable tool registers, available usually in blocks of 16; that is, a control may have 16, 32, 64, or more registers. Prior to running the program, the operator has to input (other terms used are "punch" or "dial") the desired compensation into the assigned offset register. This compensation is usually the radius of the tool.

2. To simplify the programming process in terms of roughing or finishing operations. Upon instruction, the operator can input in the offset register a radius that is larger than the actual radius of the tool used. The amount of the difference will represent the stock left on the part after the current pass.

3. To allow the use of a cutter of a different diameter than specified should a specific size become unavailable due to breakage, etc.

4. To define the direction of the compensation, to the right or to the left of the part.

5. To define the direction of the machining following the tool compensation block, i.e., conventional or climb milling.

The basic functions described above are by and large the same for most modern CNC systems. The tape block format may change, however, from one system to another. If a programmer acquires a thorough knowledge and understanding of the compensation process in one system, there should be no difficulty in switching this knowledge to a somewhat different format. Having looked at tool compensation in general terms, we will study several programs to bring the discussion to a more specific level.

7.1.1 Cutter Diameter Compensation Left—G41 (Example in Inches)

The compensation feature will position the tool, prior to machining, to the left of the part, by a distance equivalent to the tool radius input in the assigned tool offset register. The part program will be written in a very simplified cutter centerline mode, as if the tool radius were zero. The dimensions to be used for cutter motions will therefore become the actual dimensions of the part drawing, which precludes the need for calculations.

The part shown in Figure 7.1 will be machined with a 0.50 inch-diameter-end mill, in conjunction with offset register No. 03 (shown in the program as D03).

7.1.1.1 The direct approach

Using this method, the tool is programmed to move directly to point 1. The implication here is that the operator must set the holding fixture on the machine table exactly to the dimensions given in the setup drawing.

```
N0010 G20 G40 G80
N0020 G92 X0 Y0 Z0
```

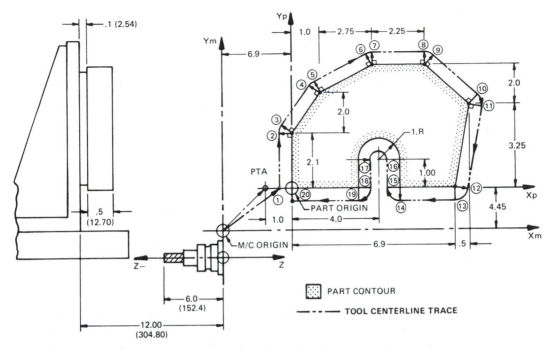

Figure 7.1 Cutter diameter compensation.

N0030 S900 M03
N0040 G91 G17 G00 G41 X6.9 Y4.45 J2.1 D03

Sequence 0040 will "rapid" the cutter from the machine origin to point 1.

- G91 represents incremental programming.
- G17 indicates that compensation takes place in the X-Y plane.
- G00 is rapid traverse.
- G41 is the code for cutter diameter compensation left.
- D03 is tool offset register No. 03, earmarked for X-Y plane programming.

As shown later, for Z-axis motion (involving tool length or "height" compensation), the same register would be designated as H03.

J is the word address indicating, as programmed, that the first motion in the X-Y plane will take place in the positive Y-direction. The control will be able to verify this intent as the program is being read ahead of the actual machine execution. Should the control detect a conflict, an alarm signal will result.

The detailed setting up of compensation using unit vectors I or J will be explained later on in this chapter. For more recent controls, compensation has been greatly simplified, and I or J are not required at all.

For a brief description, the cutter diameter compensation left (G41) or right (G42) is a displacement of the cutter center, by an amount usually equal to its radius, perpendicular, respectively, to the left or right to the vectorial resultant of the programmed vectorial components I and J. This vectorial resultant is nothing else but the hypotenuse of a right-angle triangle whose rise is J and whose run is I. In this case, I is not programmed and is therefore considered to be zero. J, as the only component, is equal to the resultant. Being positive, it points upwards. The cutter diameter compensation left, called in sequence 0040, will displace the cutter 90° to the left, as shown in Figure 7.1.

N0050 G18 G00 Z-5.3	Rapid down to 0.1 inch clearance above part surface
N0060 G18 G01 Z-0.65 F2.0	Feed down to cutting depth

The next block will have to specify a positive Y-motion, as required by sequence N0040.

7.1.1.2 The indirect approach

This method is used by experienced programmers and is far superior to the preceding one. As described previously in Chapter 6, the operator has the choice to position the holding fixture at any convenient location on the machine table, as long as it is set parallel to the machine axes. We arbitrarily select a point A, such as 1 inch away from the part, as shown in Figure 7.1. This point will be reached using G45 tool offsets. The preceding program would now read:

```
N0010 G20 G40 G80
N0020 G92 X0 Y0 Z0
N0030 S900 M03
N0032 G45 G00 X0 D01
N0034 G45 Y0 D02
N0040 G91 G17 G01 G41 X1.0 Y0 J2.1 D03
N0050 G18 G00 Z-5.3
N0060 G18 G01 Z-0.65 F2.0
```

Sequences 0010 to 0030 and 0050 to 0060 are unchanged, as is the rest of the program. The difference consists of sequences 0032 and 0034, where the operator will have to input, prior to the start of machining, the location of point A. If we assume that the part shown in Figure 7.1 has been located on the machine and its "part origin" measured at $X6.9$ and $Y4.45$, the operator will input 6.9 inches − 1.0 inch = 5.9 inches in register D01 and 4.45 in register D02. The use of two blocks for this procedure gives the operator additional flexibility.

From here on, the program will be identical for both direct and indirect approaches.

N0070 G17 G01 Y2.1 F3.0 M08	Motion from point 1 to point 2

Most compensation systems operate in two ways, depending on their executive software. The older systems require an additional code, G39, whose purpose is to rotate the tool centerpoint, as shown in the transition from points 2 to 3, 4 to 5, 6 to 7, etc., on Figure 7.1. This occurs in conjunction with a change in direction of the linear interpolation, and it is required to bring the tool radius perpendicular to the new surface to be machined. As illustrated in Chapter 5, regardless of the programming method, the cutter radius will be perpendicular at all times to the surface being machined (which in its turn is permanently "tangent" to the cutter "circle"). In the newer systems, as indicated by their manuals, the CNC software will take care of this problem by "automatically" bringing the cutter center to the intermediate position where the cutter "circle" will be "tangent" to both directions, the current one being machined, and the next one in line. For the present program, G39 will be used, as illustrated in Figure 7.2.

At the end of block 0070, the tool radius is perpendicular to part surface 1-2, between points 1 and 2. Prior to programming surface 3-4, we require a tape block to rotate the tool at point 2, so that its radius becomes perpendicular to the next surface, 3-4. It can be seen in Figure 7.2a that should this rotation not occur, the new surface will be cut to the right of the "true" line 3-4 and the part will end up smaller and scrapped.

N0080 G39 I1.0 J2.0

This tape block performs the reorientation of the tool radius. G39 is a preprogrammed subroutine, stored in control memory. The values for I and J have been selected as follows: $I = 1.0$, which corresponds to the X-displacement from point 3 to point 4 as seen in Figure 7.1. $J = 2.0$ is, in the same sketch, the corresponding Y displacement.

I and J can be interpreted as the two component vectors whose resultant is line 3-4, or as the two sides of a right-angle triangle whose hypotenuse is line 3-4. The two definitions are geometrically identical. As G41 called in sequence 0040 for compensation "left," our tool radius will now be oriented 90° to the new line 3-4, defined by I and J, as shown in N0080.

What the CNC minicomputer calculates is the slope of line 3-4, given by its rise J, divided by the run I. As the result is a ratio, we could have programmed in N0080 any pair of values for I and J whose ratio is the same. We could have had, for instance:

N0080 G39 I10 J20 or
N0080 G39 I100 J200 or
N0080 G39 I50 J100, etc.

However, it is recommended as much as possible to use pairs of values, as found on Figure 7.1, to simplify possible debugging. The G39 vector setting code must be programmed prior to each change of direction.

As seen in Figure 7.2b, illustrated for a 6M control, the motion programmed

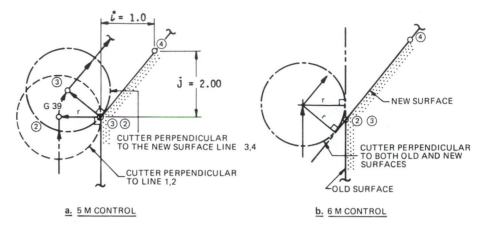

Figure 7.2 Tool radius vector setting.

in sequence 0070 will bring the cutter center to a position where the tool radius, while still perpendicular to the "old" surface, becomes perpendicular to the "new" surface as well. The software of the newer controls will automatically perform calculations and generate motions without the use of G39, or the extensive calculations from Chapter 5.

Continuing the program,

N0090 G01 X1.0 Y2.0	From point 3 to point 4
N0100 G39 I2.75 J1.15	Rotation around point 4
N0110 G01 X2.75 Y1.15	From point 5 to point 6
N0120 G39 I2.25	Rotation around point 6

As line 7-8 is parallel to the X-axis, sequence 0120 requires no J and the next one, 0130, requires no Y.

N0130 G01 X2.25	From point 7 to point 8
N0140 G39 I1.4 J-2.0	Rotation around point 8
N0150 G01 X1.4 Y-2.0	From point 9 to point 10
N0160 G39 I-0.5 J-3.25	Rotation around point 10
N0170 G01 X-0.5 Y-3.25	From point 11 to point 12
N0180 G39 I-0.9	Rotation around point 12
N0190 G01 X-0.9	From point 13 to point 14
N0200 G39 J1.0	Rotation around point 14
N0210 G01 Y1.0	From point 15 to point 16
N0220 G03 X-2.0 I-1.0	Circular interpolation

G02 or G03 will gradually rotate the tool radius as it machines the part surface; therefore, no G39 is required at either point 16 or 17.

N0230 G01 Y-1.0	From point 17 to point 18
N0240 G39 I-3.0	Rotation around point 18

N0250 G01 X-3.0	From point 19 to point 20
N0260 G18 G00 Z0.65	Raise tool to 0.1-inch clearance
N0270 G17 G40 X-1.0 Y0	Cancel cutter diameter compensation
N0280 G28 Z0	Return Z "home"
N0290 G28 X0 Y0	Return X and Y
N0300 M30	End

The process described above saves calculation time and eliminates calculation errors.

7.1.2 Cutter Diameter Compensation Left (Metric Example)

The sample part shown in Figure 7.3 is dimensioned in metric. As an observation, dimensions on metric blueprints may be shown without trailing zeros, as their dimensional tolerances are not normally related to the number of zeros after the decimal point.

The only significant change is the G21, replacing G20, which indicates to the control that the following dimensional values are given and should be reflected in

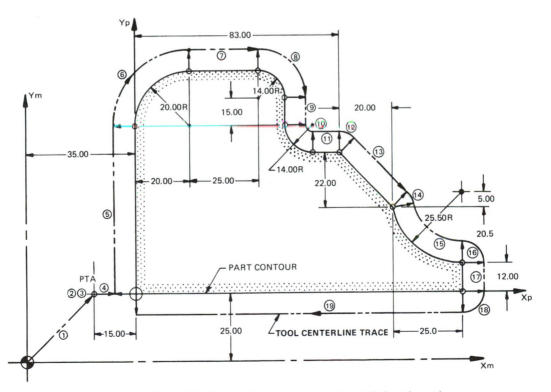

Figure 7.3   Cutter diameter compensation—left (metric part).

metric dimensions. The inch/metric is usually programmed G20/G21 on some controls, and G70/G71 on others. When a CNC system works only in metric or only in inches, the above codes are not required.

N0010 G21 G40 G80 G91

As G41 or G42 set up a cutter diameter compensation condition, G40 is the code that cancels it. An uncanceled compensation can generate a lot of problems, therefore, G40 is always used in the first block as a safety feature. G80 performs the same function for canned cycles, to be discussed in detail in Chapter 9.

```
N0020 G92 X0 Y0 Z0
N0030 S800 M03
N0040 G45 G00 X0 D01            Rapid to point A using G45
N0050 G45 Y0 D02
N0060 G18 Z-302.0               Rapid to clearance
N0070 G18 G01 Z-16.0 F10.0      Feed to cutting level
N0080 G91 G17 G01 G41 X15.0 Y0 J68.5 D03
```

For a control without the decimal point programming feature, the tape word X15.0 would be programmed X1500 for an X5.2 tape format (which means that the largest programmable dimensional word on the respective system would be X99999.99, or 5 digits before and 2 after the decimal point—a metric value representing just over 328 ft.). For a control of this nature, it may be helpful to mark the dimension 15 as 15.00, which allows for its direct transposition into the program, without the decimal point, of course, but with the correct number of trailing zeros.

```
N0090 G17 G01 Y68.5 F2.5 M08      Straight-line milling of zone 5
```

As a point of interest, one of the few applications of the "plane" codes G17, G18, and G19 occurs in cutter diameter compensation. As the compensation takes place in the X-Y plane (G17), unless the programmed Z-motion is preceded by a G18, the control will assume that the departure from the X-Y plane is accidental and the software will generate the appropriate error message.

```
N0100 G02 X20.0 Y20.0 I20.0    Circular interpolation, zone 6
N0110 G01 X25.0                Linear interpolation, zone 7
N0120 G02 X14.0 Y-14.0 J-14.0  Zone 8
N0130 G01 Y-15.0               Zone 9
N0140 G03 X14.0 Y-14.0 I14.0   Zone 10
N0150 G01 X10.0                Zone 11
N0160 G39 I20.0 J-22.0         Rotation of tool, zone 12
N0170 G01 X20.0 Y-22.0         Zone 13
```

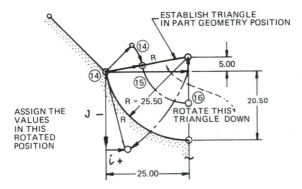

Figure 7.4 Unit vectors at rotated position.

For the cutter repositioning at zone 14 (Figure 7.4), it may be helpful to sketch an enlarged view of the part geometry in zone 15.

The cutter radius must rotate as shown at point 14 from an upper position perpendicular to zone 13, to a lower position perpendicular to zone 15. At this lower position, the cutter radius must be collinear with the radius R of zone 15. This rotated position of the tool radius must be at 90° to the left of the hypotenuse (R) of the right-angle triangle whose sides are the I and J programmed with G39.

- Draw a line from point 14 to the center of arc (R)
- From point 14 draw a horizontal line (dimension 25) and a vertical one (dimension 5)
- The three lengths, R, 25, and 5 form a right-angle triangle
- At this stage, R is collinear with the rotated position of the tool radius
- Rotate the whole triangle as shown, 90° clockwise. R is now perpendicular to the short arrow representing the cutter radius. Looking in direction of the arrow of R, the cutter radius arrow points to the left, according to G41. Therefore, R is in correct position, and the triangle side parallel with the X-axis is I (shown on Figure 7.4 as i, positive), while the other side is J (shown as j, negative).
- The rotation of the triangle has not changed the values of its sides, it has only helped to assign correct signs.

Accordingly,

    N0180 G39 I5.0 J-25.0   Rotation, zone 14
    N0190 G03 X25.0 Y-20.5 I25.0 J5.0   Zone 15

It should be observed that as G39 precedes a circular interpolation, its I and J

values are different from the ones in the subsequent motion statements, and not the same, as we had observed repeatedly in the previous program.

N0200 G39 J-12.0	Rotation, zone 16
N0210 G01 Y-12.0	Zone 17
N0220 G39 I-20.0	Rotation, zone 18
N0230 G01 X-128.0	Zone 19
N0240 G18 G00 Z16.0	Tool up
N0250 G17 G40 X-15.0 M05	Cancel compensation
N0260 G28 Z0 M09	
N0270 G28 X0 Y0	
N0280 M30	

As a closing comment, it is always advisable to set up and cancel compensation in conjunction with a straight-line motion.

Most modern controls allow simultaneous return "home" on three axes. Unless your specific control completes the Z-motion prior to the start of the X-Y, in conjunction with a G28, it is recommended to program as in the above example, to assure that the tool is out of the way prior to any side motion.

7.1.3 Cutter Diameter Compensation Right—G42

For simplicity, the tool length and part setup will be assumed the same as for the two preceding programs (see Figure 7.5).

N0010 G20 G40 G80 G91	
N0020 G92 X0 Y0 Z0	
N0030 S900 M03	
N0040 G00 G45 X0 D01	
N0050 G45 Y0 D02	To point A
N0060 G18 Z-5.3	
N0070 G18 G01 Z-0.65 F3.0	
N0080 G17 G01 G42 X0 Y1.5 I5.0 D03	
N0090 G01 X5.0 F3.5 M08	Zone 8
N0100 G39 I1.0 J1.5	9
N0110 G01 X1.0 Y1.5	10
N0120 G39 I-1.0 J1.5	11
N0130 G01 X-1.0 Y1.5	12
N0140 G39 I-2.5 J1.0	13
N0150 G01 X-2.5 Y1.0	14
N0160 G39 I-2.5	15
N0170 G01 X-2.5	16
N0180 G39 J-4.0	17
N0190 G01 Y-4.0	18
N0200 G18 G00 Z0.65 M09	

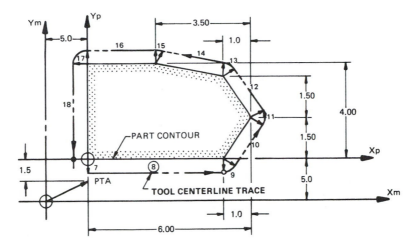

Figure 7.5   Cutter diameter compensation right—G42.

N0210 G17 G40 Y-1.5 M05
N0220 G28 Z0
N0230 G28 X0 Y0
N0240 M30

7.2 TOOL NOSE RADIUS COMPENSATION

In principle, the turning center tool nose radius is no different from the milling cutter radius. In fact, though, it is much smaller, and it does not rotate. Accordingly, we must look at a "standard tool nose vector." A vector is a quantity defined by its direction, or orientation, and its magnitude, or size.

The "direction" is usually shown as an arrow from the center of the TNR to the intersecting point of the two cutting lines of the tool.

The "magnitude" is equal to the tool nose radius when the compensation mode is activated. It is lengthened to the intersection of the two cutting lines when compensation is canceled.

In milling, the magnitude ranged from zero, i.e., cutter center, to the cutter radius, i.e., the cutter edge.

The vector of a standard tool nose is viewed from behind the tool during turning, in order to correctly assign the two directions, left or right. For example, the Fanuc 6T control provides eight different types of tool geometries, as shown in Figure 7.6. Depending on which cutter geometry suits a particular application, the programmer must also select this standard tool nose number together with its corresponding code.

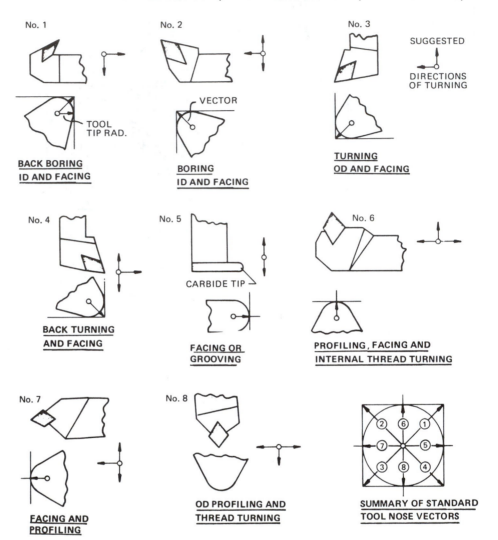

Figure 7.6   Standard tool nose tools.

The TNR and the number of the standard tool nose must be input into the respective tool offset register prior to running the program, as shown in Table 7.1.

Turning centers usually have 16 or 32 programmable tool offset registers. These can be shown in the program in the tool word as follows:

N25 T0212

The first two digits following the "*T*" refer to the tool turret position. The second pair of two digits represent the tool offset register number.

TABLE 7.1 TOOL OFFSET SHEET

Offset register number	Compensations		Tool nose radius R	Standard tool number
	X	Z		
12	-0.006	0.009	0.031	4

TNR compensation left, G41, when programmed, will lead to the position shown in Figure 7.7:

- The tool tip is to the left of the surface already machined.
- The direction of the tool parallel to the turned surface is to the left of the vector perpendicular to the machine surface.

This compensation condition is maintained by the control unit until a compensation cancel code, G40, is programmed. The G41 code is most often used for internal turning and facing.

TNR compensation right, G42, when programmed, will position the tool tip as shown in Figure 7.8:

- The tool tip is to the right of the surface already machined, as we look at the turned surface, along the vector perpendicular to it.
- The direction of the tool parallel to the turned surface is to the right of the vector perpendicular to the machined surface, as we look along it, from its tail to its tip.

TNR compensation right is used for contour turning of outside diameters.

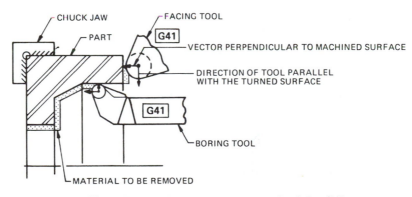

Figure 7.7 Tool nose radius compensation left—G41.

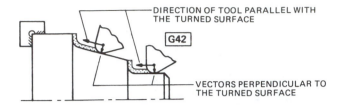

Figure 7.8   Tool nose radius compensation right—G42.

In both cases, left and right, the control will maintain the vector perpendicular to the turned surface, regardless of the type of contour.

7.2.1 Setting Up Tool Nose Radius Compensation

The program line that activates the compensation is called the startup line or tape block. It can only be programmed in G00 or G01 modes. Figure 7.9 illustrates a typical tool compensation programming situation for both G41 and G42.

Assuming the tool to be at the start point "A" for both cases, we can ignore the tool dimensions during programming, since we are programming the part, not the tool. The tool dimensions will have to be input in tool offset register No. 7 prior to turning the part.

N0010 G00 X2.5 Z0.5 T0707	Rapid to clear part
N0020 G01 G41 Z0 F0.05	Compensation at "B"

or

N0010 G00 X4.0 Z-2.4 T0707	Rapid to clear part
N0020 G01 G42 X3.75 F0.05	Compensation at "C"

In both cases the tool compensation was set from a short, safe distance away from the part, to save machining time. As well, both applications show machining of surfaces parallel to the primary axes of the coordinate system.

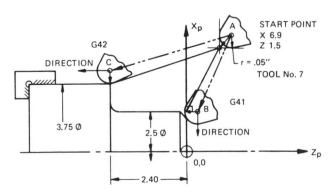

Figure 7.9   Startup in tool compensation mode.

7.2.2 Canceling the Compensation—G40

Canceling the compensation—G40 is the reverse process of setting the tool compensation. We shall assume that the turning was finished at points "*B*", respectively "*C*," and it is desired to cancel the tool compensation condition. It is good practice to move the tool to a safe location away from the part before the G40 code is programmed.

N0030 G01 Z0.5 F35.0	Tool away from part
N0040 G40 G00 X6.9 Z1.5	Cancel compensation

These two blocks could also be written as follows:

N0030 G01 X4.0 F35.0	Tool away from part
N0040 G40 G00 X6.9 Z1.5	Cancel compensation

This safe method of programming should be used until one becomes reasonably certain that the tool will move away from the part in the compensation cancel block. The two above versions could be programmed, omitting sequence N0030.

When compensation is canceled directly off the part surface, the control provides a special programming feature for tapered surfaces, using unit vectors as shown in Figure 7.10.

Assume the last tool motion programmed to be from point *A* to point *B*.

N0050 Z-1.75

At this point, the tool tip center is under compensation control, and its radius is tangent to both cylindrical and tapered surfaces at the end of the motion.

If at this point the tool compensation is canceled, prior to returning to point 0, we must program in the same block an "I" and a "K" in order to prevent the tool center point from jumping in line (vertically) with point *B*. This would result

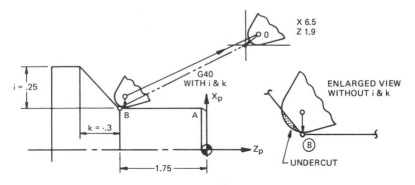

Figure 7.10 Tool compensation using vectors i and k.

in an undercutting position, as illustrated in Figure 7.10. The correct programming procedure is shown below in sequence N0060.

N0060 G40 G00 X6.5 Z1.9 I0.25 K-0.3 T0700

Bear in mind also that while the compensation feature is on, each and every line of program must contain a motion. No tape blocks should be inserted that contain only *M*, *S*, *G* codes, or time delays, because the result will be an undercutting condition.

7.2.3 Tool Nose Radius Compensation Left—G41

As mentioned earlier, TNR compensation left—G41 is normally used for internal turning. In Figure 7.11, the tool will rapid to a safe clearance of 0.5 inch. The compensation will be set up in conjunction with a G01 motion.

To simplify the program and keep it to a manageable size, it shall be assumed that the part was already rough-turned.

N0010 M41	2nd gear range
N0020 G50 X0 Y0 Z0	"Zero" machine in home position
N0030 G99	Feed in ipr
N0040 G00 X-5.3 Z-13.0	Rapid to point 0
N0050 G50 X8.0 Z3.0	Transfer coordinates to part
N0060 T0202 S1200 M03	Index tool
N0070 G00 X3.6 Z0.5	Rapid to point 1
N0080 G01 G41 X3.9 Z0.1 F0.06	Set compensation to point 2
N0090 X3.5 Z-0.1 F0.005 M08	To point 3
N0100 Z-0.75	To point 4
N0110 X2.634 Z-1.5	To point 5

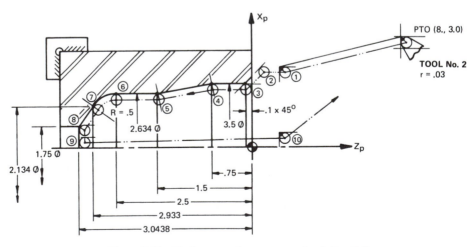

Figure 7.11   Tool nose radius compensation left—G41.

N0120 Z-2.5	To point 6
N0130 G03 X2.134 Z-2.933 R0.5	Circular interpolation to point 7
N0140 G01 X1.75 Z-3.0438	To point 8
N0150 X1.5	To point 9
N0160 G00 G40 X1.4 Z0.5	To point 10

The X1.5 in sequence N0150 moves the tool point a safe distance away from the part prior to canceling the tool compensation. The alternate method is to program a negative K value, to prevent the tool from moving in the negative X-direction at point 8, which would result in an undercut.

In the second method, sequence N0150 becomes redundant, and the balance of the program is shown below:

N0160 G00 G40 X1.4 X0.5 K-5.0	Cancel compensation
N0170 X8.0 Z3.0 T0200	Return to point 0
N0180 G50 X-5.3 Z-13.0 M09	Coordinates back to machine
N0190 G28 X0 Z-13.0 M05	Turret slide to "home"
N0200 M30	End of program

Tool offset register No. 02 will have to be set prior to running the program.

7.2.4 Tool Nose Radius Compensation Right—G42

Tool nose radius compensation right—G42 is illustrated in the program below, written for the part shown in Figure 7.12.

N0010 M41	Second gear range
N0020 G50 X0 Z0	"Zero" machine in home position
N0030 G99	Feed in ipr
N0040 G00 X-4.0 Z-16.5	Rapid to point 0
N0050 G50 X8.0 Z5.0	Transfer coordinates to part

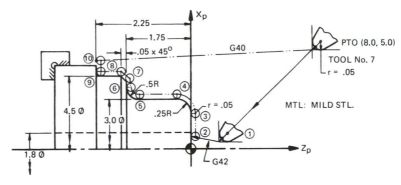

Figure 7.12 Tool nose radius compensation right—G42.

N0060 T0707 S1200 M03	Index tool
N0070 G00 X1.6 Z0.2	Rapid to point 1
N0080 G96 S680	Constant surface speed
N0090 G01 G42 X1.7 Z0 F0.05	Set compensation, motion to point 2
N0100 X2.5 F0.005 M08	Finish turn face to point 3
N0110 G03 X3.0 Z-0.25 R0.25	Contour to point 4
N0120 G01 Z-1.25	Turn 3.0-inch diameter to point 5
N0130 G02 X4.0 Z-1.75 R0.5	Contour to point 6
N0140 G01 X4.4	Face to point 7
N0150 X4.5 Z-1.8	Turn chamfer to point 8
N0160 Z-2.25	Turn 4.5-inch diameter to point 9
N0170 X5.1	Face, exit to point 10
N0180 G00 G40 X8.0 Z5.0 T0700	Cancel compensation
N0190 G50 X-4.0 Z-16.5	Coordinates back to machine
N0200 G28 X0 Z0 M05	Turret slide to "home"
N0210 M30	End of program

Prior to running this program, we would have to insert the tool tip radius of 0.05 inch and the standard tool No. 3 into tool offset register No. 07, in manual data input (MDI) mode.

EXERCISES

7.1. What is the essential concept of cutter diameter compensation?

7.2. What are the major advantages of cutter diameter compensation?

7.3. How many types of cutter diameter compensation are there, and what functions are attached to each type?

7.4. Does this code actually displace the tool by the value of the diameter?

7.5. Explain each word in the program line:

N50 G17 G91 G00 G41 J2.0 D05

7.6. How does J, in the preceding line of program, or I and J in general, affect the displacement of the cutter center?

7.7. What is the major difference in setting up tool nose radius compensation in a turning job by comparison to a two-axis milling job?

7.8. Why should compensation be cancelled, when should it be cancelled, and what precautions are indicated?

7.9. Write programs for the parts in Figures 5.13, 5.14, and 5.15, pp. 90–91, using cutter diameter and respectively tool nose radius compensation.

8

Tool Length
Compensation

Tool length compensation is a code found on a number of machining centers equipped with more recent controls. It was designed with a specific purpose in mind, unlike the multipurpose tool offset discussed in Chapter 6.

Tool length compensation, as its name indicates, is used to compensate for tool length differences. It can perform this function in two directions:

1. G43, away from the part, or upwards in a vertical machine
2. G44, toward the part, or downwards if the machine is vertical.

As in other compensations previously discussed, cancelation is provided through a tool length compensation cancelation code, G49.

The major advantage of this feature is that it provides the programmer with the ability to write a complete program without knowing exactly the length of the tools to be used. The alternative is a set of very accurate calculations involving the distance from the quill face to the table; the dimensions of holding fixtures, spacers, risers, material thickness; the exact length of the holder-tool combination; and all the "Z" motions. These calculations would have to be performed by the programmer, rigorously implemented by the operator, and modified every time anything changes.

Tool length compensation introduces an elastic link in this chain of dimensions. The programmer can now write the entire program, including all the vertical motions, by using an imaginary tool length. This may be shorter or longer than all the tools the operator is likely to load in the automatic tool changer.

8.1 TOOL LENGTH COMPENSATION AWAY FROM THE PART—G43

In the case of a vertical machine, it is probably easier to visualize G43 as the tool length compensation "upwards," as this motion gets it away from the part surface.

The principle is the following: The programmer will assume that all the tools to be used are zero in length, i.e., he or she will program the quill face (adjusted for any gap that may exist on a particular machine between the quill face and the back of the tool holder). It should be noted that the reference to "tools" always indicates the holder-cutting tool combination.

Accordingly, should the distance from the quill face to the surface to be attained be, e.g., 15 inches, the appropriate program line would read:

```
N0010 M06 T01
N0020 G91 G00 G43 Z-15.0 H01
```

In tool offset register No. H01, prior to running the program, the operator will have inserted, in MDI mode, the true measured length of tool No. 1 (T01 from sequence 0010), e.g., 11 inches.

Consequently, the actual rapid motion of the spindle toward the part will be:

$$15.0 \text{ inches} - 11.0 \text{ inches} = 4.0 \text{ inches}$$

The major advantage of this approach is the time-saving for the programmer who no longer has to worry about tool lengths throughout the program. The increase in efficiency at the machine is also appreciable, since the setup will no longer have to be achieved to a set of given dimensions. The operator now makes the setup, measures it, and records the measurements. Should tool T01 have to be replaced later on by a reground version, different in length, the operator only has to update the information in tool offset register H01. The program remains unchanged, and the operator need not access it in memory should the lengths of the tools vary.

The potential pitfall is of course that the "cutterless" motions programmed are perforce fairly long, and should the operator overlook inputting the tool length value, a serious crash would result in the Z-axis.

It can be seen that in the absence of a tool length compensating feature, the programmer would have to know exact tool lengths and program exact Z-motions. Any change in the length of any tool would result in a requirement for a program change. The compromise, in the absence of G43, is the use of the tool offsets, such as G45, discussed in Chapter 6.

8.2 TOOL LENGTH COMPENSATION TOWARD THE PART—G44

The other version available with this feature is the tool length compensation toward the part, or "downwards" in a vertical machining center.

The principle here is the opposite to the previous "zero-length" assumption. The programer will select an identical very large value for the length of all the tools to be used in the program. A suitable value is the nearest rounded-off value to the distance between the retracted quill face and the part surface, as shown in the following example.

Let the known (measured) distance between the table surface and the retracted quill face be 32.4000 inches. Assume that the measured distance between the table surface and the part surface to be machined is 3.2500 inches, including part thickness and fixturing. The calculated distance between the retracted quill and the part surface will be:

$$32.4000 \text{ inches } - 3.2500 \text{ inches } = 29.1500 \text{ inches}$$

The amount of clearance required by the downward motion of the automatic tool changer is, e.g., 6.5000 inches.

$$29.1500 \text{ inches } - 6.5000 = 22.6500 \text{ inches}$$

The programmer can assume in the program that all the tools to be used are, e.g., 20.0000 inches long, and all the Z-motions will be programmed accordingly.

As the tools are quite "long," these Z-motions will be fairly "short."

The operator will set up the tools in accordance with the instruction sheet, will measure their exact length, will subtract the measured length from the value 20.0000 inches, supplied on the sheet, and will insert each difference in the respective tool offset register, as outlined by the Tool and Offset Sheet. The control will add these "differences" throughout the program, to the "short" motions specified by the programmer, thereby compensating for the difference in length between the actual values and the imaginary 20-inch tool.

Although this method entails an additional operation on the part of the operator, i.e., subtracting the length measured from the imaginary length supplied with the program, from a safety standpoint it is preferable to the other one. If the operator forgets to insert a value in the appropriate tool register, the motion will be short, but no crash will result.

Either length compensation method may be used, if available on the control. Once a method is adopted, there should be no deviations from it as they lead to confusion, and expensive crashes could result.

Tool length compensation may be used in conjunction with two types of motion:

- X-Y plane motions, as in milling
- Z-axis motions, as in drilling, tapping, or boring.

Where four-axis programming is involved, the control manual should be carefully checked as tool length compensation may be inoperable, or subject to a set of limitations.

In *X-Y* operations, it is advisable to set up the tool compensation in combination with a rapid (G00) motion to a safe clearance above the part. The balance of operations will take place in feed (linear or circular, as applicable).

In Z-axis operations, the compensation may be set up in conjunction with a rapid motion, to the "Initial level" of the subsequent canned cycle operation (see Chapter 9). While the tool would still continue to the "*R*" level in rapid, this method would allow return to the "Initial level" without affecting the compensation condition.

8.3 TOOL LENGTH COMPENSATION CANCELATION—G49

Once the *X-Y* operations have been completed and the tool returned to the clearance location, a rapid return to start in *Z*, with a G49, will cancel the compensation. The same will apply to a succession of Z-axis operations, once the tool was returned in rapid to the "Initial level."

It is advised that the cancelation of tool length compensation take place in the same length of motion (but the opposite direction) as the initiating motion in G43 or G44.

As are many other CNC features, tool length compensation is particularly appreciated by people who programmed and made parts on NC machines before it became available.

EXERCISES

8.1. What is tool length compensation used for?

8.2. List and explain the tool length compensation codes.

8.3. What is the major advantage of this feature?

8.4. What is the tool length compensation programming principle?

8.5. Is there an axis restriction in setting up tool length compensation?

8.6. Discuss cancelation of tool length compensation.

8.7. Write programs for the parts in Figures 5.13 and 5.14, pp. 90–91, using tool length compensation codes.

9

Canned Cycles

Canned cycles may be defined as a set of preprogrammed instructions stored away in computer memory. The word "canned" has probably been borrowed from canned goods, which one usually stores away for later use. Because the instructions represented a set of routinelike repetitive patterns, the word "cycle" was found to best express what was taking place. These canned cycles are filed away under a G-code address. To a large extent the G-codes are standardized.

As an example (see Figure 9.1), a G84 code, usually representing right-hand tapping in a machining center, will consist of the following steps:

- Clockwise rotating of the tap at the correct rpm
- Rapid advancing of the tap to a set clearance from a predrilled hole (R-level)
- Feeding the rotating tap to a set depth at a rate of one thread pitch per revolution (or a set displacement in mm/rev. or in in./rev.)
- Reversing both feed and spindle rotation until the tap reaches the R-level
- Returning spindle rotation to its original clockwise direction

The spindle is now ready to rapidly return "home" for a tool change, relocate to another position or carry out any other instruction of the program.

Similar motion patterns are available for drilling, boring, turning, threading, etc.

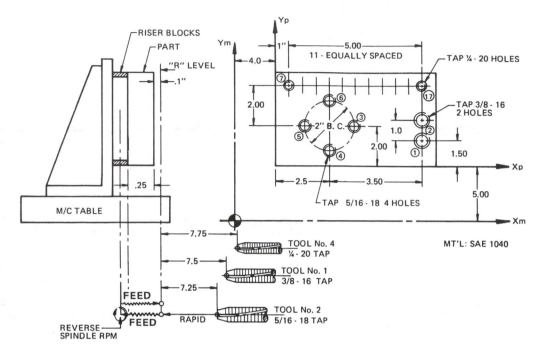

Figure 9.1 Tapping canned cycle G84.

Multiple Repetitive Cycles or Automatic Repeat Cycles.   These represent the generating of complex profiling patterns in stock removal. They are very similar to canned cycles in that they are preprogrammed and prestored. Together the canned cycles and the multiple repetitive cycles are known as Fixed canned cycles.

Variable Canned Cycles or Subroutines.   These emerged because of the increase in the number and complexity of prestored canned cycles and multiple repetitive cycles. This led to increases in the size of computer size and price, combined with slowdowns in the speed of processing.

The variable canned cycles allow the programmer to write a canned cycle to suit a particular requirement or application. A larger portion of the memory will now be available to handle a set of specific instructions, perform calculations, manipulate variables, and carry out other assignments at a far higher rate of speed.

9.1 FIXED CANNED CYCLE PROGRAMMING

9.1.1 Machining Centers, Vertical or Horizontal

As usual, the tool is always considered to be traveling in relation to a fixed part regardless of the fact that in one or more directions, the actual motion may be carried out by the machine table.

The Right Hand Coordinate System.   With positive "*Z*" pointing toward the quill, this system will define the three axes of linear motion for a given machine, be it vertical or horizontal.

In the following discussions, we have selected a horizontal machine for our programs. We have also used decimal point programming, as its use is becoming more widespread. Thus a motion of 3.5 inches will be shown as X3.5 in lieu of the more conventional X35000. A motion of 2 inches would be shown then as Y2.0, or simply Y2., in lieu of Y20000.

The *general format* of the fixed canned cycle is the following:

$$N\_G\_G\_X\_Y\_Z\_ R\_Q\_P\_F\_L\_$$

where (see Figure 9.2),

> N is the sequence number of the tape block
> G is the respective canned cycle (e.g., G81, G73, G76, etc.)
> The second G may be G98 or G99, to be discussed later
> X and Y are the coordinates of the location to be machined
> Z is the level attained by the tool in feed mode
> R is the level reached by the tool at the end of its rapid motion
> Q is the depth of cut in feed in peck drilling or the lateral shift in boring canned cycles
> P is the dwell in spot facing or boring operations
> F is the feed code
> L is the number of repeats, if applicable

(Codes not shown in Figure 9.2 will be discussed in detail later in this chapter.)

The following observations apply to canned cycles for machining centers in general:

- The tool motion starts in canned cycle from an Initial Level.
- The tool will move in Rapid mode to the programmed *X-Y* location.
- The quill will move in Rapid mode to the *R* (for Rapid) Level.
- The tool will feed to the *Z*-Level, measured from Initial in absolute mode (G90) and from *R* in incremental mode (G91).
- If the tool is already on location, there is no need to program *X* and *Y*.
- G99 will return the tool point to *R* level; G98 will return the tool point to the Initial level.
- A canned cycle is in effect until canceled (or replaced by another canned cycle). Accordingly, subsequent locations can be machined by simply programming *X* and *Y* in subsequent blocks. *If you do not intend* to perform the canned cycle at the next location, you must cancel it first.

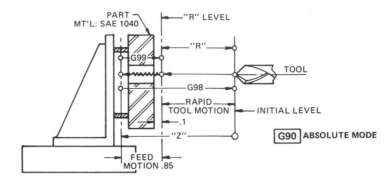

CANNED CYCLE IN ABSOLUTE MODE

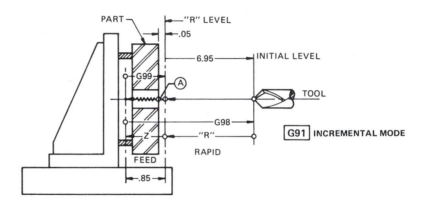

CANNED CYCLE IN INCREMENTAL MODE

Figure 9.2 Canned cycle in (a) absolute mode and in (b) incremental mode.

- A canned cycle can be programmed in either incremental or absolute mode.
- Since *X-Y* motions are strictly for positioning and the programmed machining patterns take place in the *Z*-axis, no canned cycles are to be used in cutter diameter compensation mode.
- In the examples shown below, tool offsets and cutter length compensation will be minimized or omitted in order to allow the reader to concentrate on the respective canned cycle process. Preset tooling will be used.
- Most systems provide numerous canned cycles as standard equipment, but not all systems use the same codes to designate the same cycles. Always use the codes provided by your system manual.

In summary, therefore, there are three levels of tool position in any canned cycle:

- The Initial level, which is the specific position of the tool at the moment the canned cycle becomes effective,
- The "*R*" for Rapid level, which is the end of the rapid quill motion, usually 0.020 to 0.100 inches (0.5 to 2.5 mm) from the part surface, and
- The "*Z*" level which is the end of the feed (metal cutting) motion.

Usually, in one tape block, the canned cycle will control all or most of the following motions:

- Positioning the tool in the *X-Y* plane (Remember: the machine may position the part, but it is the tool motion that we program),
- Rapid motion to the "*R*" level,
- Controlled feed motion to the "*Z*" level,
- Commanded motions at "*Z*" level, such as dwell, spindle stop, spindle reverse, spindle orientation, spindle lateral shift, etc.,
- Rapid or feed return to "*R*" level, as applicable, and
- Rapid return to INITIAL level.

9.1.1.1 Canned cycle cancelation G80

This code is normally used in three situations:

- When different programs are run consecutively on the same machine, it is used before the first motion statement, in conjunction with other codes as a safety feature, to cancel potential leftover canned cycles from a previous program. On some controls, a previously used canned cycle may remain active even if the control is reset or the power is turned off.

N0010 G20 G40 G80 G49 G91 G00

- When a canned cycle is canceled, as it should be, as soon as it is no longer needed in the program.

N0040 G80

- When it is required to reposition the spindle without any machining taking place at that location (such as an intermediary position in a change of direction).

N0090 G80 X3.0 Y4.5

(Inch programming, shown above as G20, is programmed on many controls as G70. Check your programming manual.)

9.1.1.2 Drilling canned cycle G81

This canned cycle will perform the drilling of one or more holes as follows: Rapid to "R" level, feed to "Z" level and rapid return to "R" (in G99) or to Initial level (in G98). See Figure 9.3.

(NOTE: Comments or explanations placed on the program lines do not form part of the program. Some controls on the market allow the programmer to place data or instructions in brackets at the end of the program line. Such data will show on the CRT but will be overlooked by the machine. Other controls allow comments on a separate line, preceded by a specific G code.)

N0010 G20 G40 G80 G91	inch programming shown as either G20 or G70, cancel cutter diameter compensation, cancel canned cycles, incremental
N0020 B180	index table 180°
N0030 M08	coolant ON
N0040 M06 T01	tool change, select tool No. 1
N0050 S1500	spindle speed 1500 rpm
N0060 G99 G81 X7.75 Y5.3125 Z-0.25 R-7.25 F3.5 M03	center drilling of hole 1, 0.2-inch deep; (note that this program does not concern itself with the tool point-to-part surface distance or tool length offsets in order to concentrate on the canned cycle programming aspects)
N0070 X-0.4375 Y0.4375	center drilling of hole 2; (observe that the control only requires the location of this hole; the canned cycle remains active. G99 from N0060 will rapid return the tool point to R level from which the next hole will be machined following repositioning to the still active Z-level)
N0080 X0.4375 Y0.4375	hole 3
N0090 X0.4375 Y-0.4375	hole 4
N0100 X2.0625 Y1.5	hole 5
N0110 X0.75	hole 6
N0120 X0.75	hole 7
N0130 Y-0.75	hole 8
N0140 G98 Y-0.75 M05	hole 9; return to Initial; spindle stop
N0150 G80 M09	cancel canned cycle; stop coolant
N0160 G28 Z0	quill returns "home"
N0170 M06 T03	tool change, select tool No. 3
N0180 S1200 M03	select spindle speed, spindle start clockwise (note that in sequence N0050 we have only programmed the spindle speed, with the actual spindle start in the next sequence; this and other apparent discrepancies are intended to show that instructions may be written in a number of ways, for increased programming flexibility)

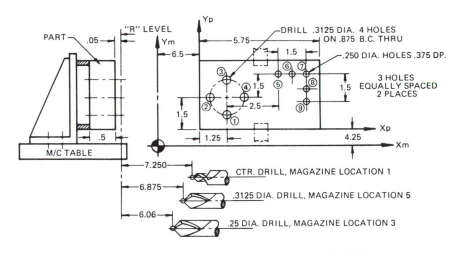

Figure 9.3 Sample part for drilling canned cycle G81.

N0190 G99 G81 Z-0.425 R-6.06 F4.0 M08

drill hole 9; the G81 canned cycle was reset with the distances corresponding to the 0.25-inch drill; the tool change has allegedly taken place at the same location therefore no X and Y are programmed in N0190, and the drilling will take place in reverse order, thus avoiding unproductive machine travel

N0200 Y0.75 — drill hole 8

N0210 Y0.75 — drill hole 7

N0220 X-0.75 — drill hole 6

N0230 G98 X-0.75 M09 — drill hole 5; return to Initial level

N0240 G80 M05 — cancel canned cycle; spindle stop

N0250 M06 T05 — tool change; select tool in magazine holder No. 5

N0260 G99 G81 X-2.0625 Y-1.5 Z-0.65 R-6.875 F4.0 — drill hole 4

N0270 X-0.4375 Y0.4375 — drill hole 3

N0280 X-0.4375 Y-0.4375 — drill hole 2

N0290 G98 X0.4375 Y-0.4375 M05 — drill hole 1; return to Initial level

N0300 G80 M09 — cancel canned cycle; coolant off

N0310 G28 X0 Y0 — return to machine origin

N0320 M30 — end of program.

The program could have been written in absolute, based on Figure 9.2.

In comparison with conventional programming, the canned cycle operation is far more cost-effective. The above canned cycle program could be made even more efficient by the use of the subroutine feature, when several operations (center drill, drill, tap, etc.) are performed at the same locations. The location pattern is placed in a subprogram known as subroutine and called as needed by the main program which sets up the machine and changes tools. Any straight-line equally spaced hole pattern program can be drastically shortened by the use of the "*L*"

word address, representing the number of repeats of the *incremental* displacement
set up in the *X-Y* plane.

N0010 . . . .   sets up the canned cycle
N0020G91 X . . . Y . . . L . . .   the above canned cycle will be
                                       repeated L times, each time at a
                                       position shifted over by X and Y
                                       from the preceding tool location,
                                       hence G91

The following application illustrates repetitive canned cycle programming (see
Figure 9.4). The part drawing shows two straight-line equally spaced hole patterns.
The upper row has 20 spaces, *X*-only motions. The overall distance is 10.0 inches,
the spacing is 0.5 inch.

The lower row has 10 spaces. The *X*-space is provided by the same overall
10.0-inch distance as above, hence an *X*-spacing of 1.0 inch. The *Y*-space is obtained
by dividing the 10 spaces into the 4.0-inch dimension, for a *Y*-spacing of 0.4 inches.

In both cases, the canned cycle will be set up for the first hole in the pattern,
while a single repetitive cycle block will take care of all the remainder.

N0010 G20 G40 G80 G91
N0020 B180
N0030 M06 T01
N0040 S2200 M03
N0050 G99 G81 X6.5 Y10.5 Z-0.25 R-8.65 F3.0 M08   This block will drill the first hole at point "A" in the
                                                   upper row
N0060 G91 X0.5 L20   The sequence of operations initiated by N0050 will be

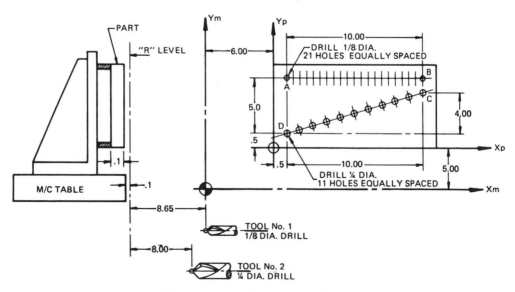

Figure 9.4 Repetitive canned cycle programming.

repeated 20 times. G91 is redundant in block 0060 since it appears in block 0010, but it has been placed here to emphasize the incremental nature of "L." The reiteration of G91 will not create any problems, and the tool will travel from hole to hole at R level

N0070 G80 M09
N0080 G28 Z0 M05
N090 M06 T02
N0100 S1600 M03
N0110 G99 G81 Y-1.0 Z-0.25 R-8.0 F3.5 M08

This block will drill the first hole of the lower row at point "C." The remaining 10 holes will be drilled in a single repetition cycle block in the next sequence

N0120 G91 X-1.0 Y-0.4 L10

The sequence of operations initiated in sequence N0110 will be repeated 10 times at successive equally spaced locations

N0130 G80 M09
N0140 G28 Z0 M05
N0150 G28 X0 Y0
N0160 B180
N0170 M30

9.1.1.3 Tapping canned cycle G84

This canned cycle will perform the tapping of one or more holes as follows: Rapid to "R" level, feed to "Z-" level, reverse direction of rotation at "Z-" level, feed to "R" level, reverse direction of rotation to original direction.

The programmed feed must be fully synchronized with the spindle speed to avoid tool breakage, and the control will ignore speed and feed overrides, as well as feed hold and single block to the end of the sequence in progress.

To avoid an unproductive lengthening of the program, we'll assume that drilling has already taken place and that the "R" level is on the safe side of the clamps. Before starting the tapping portion of the program at, e.g., sequence N0100, we have to set tapping speeds and calculate tapping feeds.

Suppose the tapping speed has been selected to be 100 rpm (S100), the tapping feed for $3/8''$-16 could be calculated as follows:

1. Sixteen tpi represent a pitch of $1/16$ or 0.0625 inch. For one revolution of the tap, the quill will advance 0.0625 inch. For 100 revolutions, the quill advance or feed will be 100 × 0.0625 or 6.25 inches. As the above 100 revolutions take place in 1 minute (rpm), the amount 6.25 represents ipm, and it is the tapping feed, F6.25.

2. It may be shorter to just divide the selected speed by the number of tpi. Therefore, $100/16 = 6.25$ and we program F6.25. Note that the programmer should use the maximum number of decimal places for maximum accuracy of the process.

3. When using an "inch" tap in a metric program, we just "soft-convert" the feed resulting from the method above to millimeters. F6.25 becomes F158.75, by multiplying 6.25 inches by the factor 25.4.

4. For metric threads in a metric program, since these are defined as the diameter times the pitch in mm, we just multiply the pitch by the selected rpm as in method 1. Using 125 rpm to tap M6x1, the feed will be 125×1 or F125.

5. For metric threads in an inch program (highly improbable, but not impossible), we simply divide the feed obtained as in method 4 by 25.4. Hence, 125/25.4 will be programmed as F4.921259, or on some recent controls as E4.921259.

The other tapped hole in our program is $\frac{5}{16}$"-18 tpi. The feed will be, given 100 rpm tapping speed: $\frac{100}{18}$ = F5.5555556. For the last hole size, at 125 rpm, a $\frac{1}{4}$"-20 will be tapped at F6.25.

N0100 S100 M03	spindle on, 100 rpm, clockwise. Coolant assumed on
N0110 G99 G91 G00 G84 X10.0 Y6.5 R-7.5 Z-0.5 F6.25	tool No. 1, assumed changed prior to seq. 0100 will rapid to the incremental dimensions X and Y, will rapid to "R," will tap to "Z,", reverse and return to "R" where the clockwise rotation will be reestablished
N0120 G98 Y1.0	tap hole 2
N0130 G80	
N0140 M05	

Assuming that tool changing and offsetting for tool No. 2 has taken place according to the data in Figure 9.1, we continue with sequence 180.

N0180 S100 M03	
N0190 G99 G84 X-2.5 Y-0.5 R-7.25 Z-0.5 F5.556	tap hole 3
N0200 X-1.0 Y-1.0	tap hole 4
N0210 X-1.0 Y1.0	tap hole 5
N0220 G98 X1.0 Y1.0	tap hole 6, return to Initial Level.

Following tool changing and offsetting for tool No. 3, we can reestablish the tapping canned cycle at hole 7, then program 10 repeats with an X pitch of 0.5 inch. Assuming sequence 260, we continue:

```
N0260 S125 M03
N0270 G99 G84 X-1.5 Y1.0 R-7.75 Z-0.5 F6.25
N0280 X0.5 L10
```

Following completion of tapping, the program may be terminated the usual way or continued for other operations.

9.1.1.4 Fine boring canned cycle G76

Fine boring canned cycle G76 code is one of a variety of boring cycles designed by the various manufacturers to cover just about any industrial application.

The canned cycle will perform the boring of one or more holes as follows: Rapid to "*R*" (point A, Figure 9.5). Feed to "*Z*" (point *B*), spindle stop and orient, shift the tool point by the programmed amount *Q* away from the hole wall, rapid return to "*R*" or Initial as programmed, and restart of spindle. The following program will be in absolute, and as usual will place its emphasis on the canned cycle it illustrates.

```
N0010 G20 G40 G80 G90 G00
N0020 T01 M06
N0030 G92 X0 Y0 Z0                    this command will zero the registers with no
                                      motion taking place

N0040 S600 M03
N0050 M08
N0060 G98 G76 X9.0 Y7.5 Z-5.9 R-4.75 Q0.1 F0.75
```

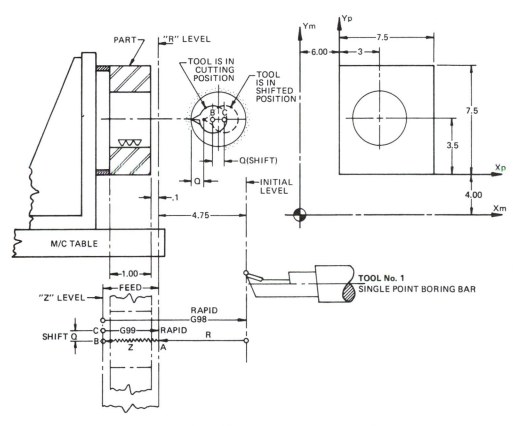

Figure 9.5 Fine boring canned cycle G76.

At 600 rpm, F0.75 will ensure 800 tool marks or ridges in 1 inch of bored surface. Should a finer surface be required, we can increase *S*, reduce *F*, or both. In a production situation, a carbide-tipped boring tool could run at speeds to 3,000 rpm and feeds to 3.5 ipm. The introduction of fine boring cycles to CNC machining centers, particularly when the spindle is temperature-controlled, has drastically reduced the use of expensive conventional jig-borers. The above boring program can of course be continued or finished to suit as necessary.

9.1.1.5 Boring canned cycle G89

Somewhat similar to the precedent, this code will spotface, bore, or counterbore. The canned cycle will perform as follows: Rapid to "*R*," feed to "*Z*," dwell at "*Z*" for a duration given by "*P*" and rapid return to "*R*" or Initial depending on whether G99 or G98 was called in the program. See Figure 9.6. One second is P100 and 3 seconds are P300, depending on the control, and decimal point programming does not usually apply to the *P* tape word. The program application below will illustrate both boring modes discussed in this chapter. Note that because of the central boss, the figure shows two "*R*" levels. See Figure 9.7.

We shall assume that the part has been supplied with the two 0.875 holes predrilled, but not bored, and the program will look as follows:

```
N0010 G20 G40 G80 G91
N0020 B180
N0030 T01 M06
N0040 S1600 M03
N0050 M08
N0060 G99 G81 X12.5 Y10.5 Z-0.25 R-7.5 F3.0    center drill hole 1
N0070 G98 X-6.0                                 center drill hole 2, return to
                                                    Initial to avoid crash on the way to hole 3

N0080 G99 Y7.5                                  center drill hole 3
N0090 X2.0 L3                                   center drill holes 4, 5, and 6
N0100 G28 Z0 M09                                retract quill to machine Z-origin
N0110 G80 M05                                   cancel canned cycle, stop spindle
N0120 T02 M06
```

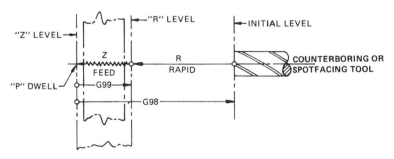

Figure 9.6 Boring canned cycle G89.

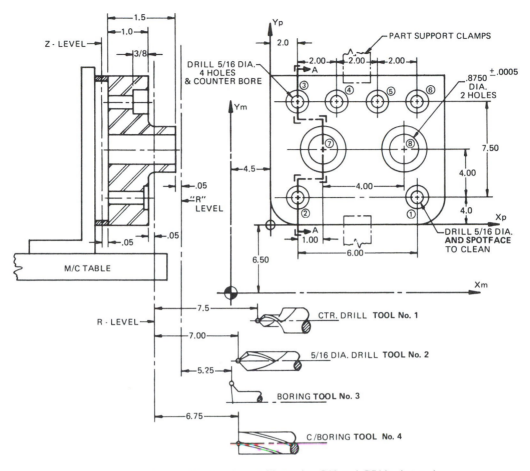

Figure 9.7   Sample part illustrating G89 and G76 boring cycles.

If there is sufficient clearance, performing a tool change on top of hole 6 will allow immediate continuation of the machining process without idle *X*- and *Y*-travel off the part and back on again.

N0130 S1500 M03	
N0140 G99 G81 R-7.0 Z-1.2 F3.5	drill hole 6
N0150 X-2.0 L2	drill holes 5 and 4
N0160 G98 X-2.0	drill hole 3, return to Initial Level
N0170 G99 Y-7.5	drill hole 2
N0180 G98 X6.0	drill hole 1
N0190 G80 M05	cancel canned cycle, spindle stop
N0200 G28 Z0 M09	

Again we have assumed that we can carry out a tool change on top of the part. As boring tools are usually longer, dimensional verification is advised.

N0210 T03 M06	
N0220 S1500 M03	
N0230 G99 G76 X-1.0 Y4.0 R-5.25 Z-1.6 Q0.1 F0.75	fine bore hole 8
N0240 G98 X-4.0	fine bore hole 7
N0250 G80 M05	
N0260 G28 Z0 M09	
N0270 T04 M06	
N0280 S500 M03	
N0290 M08	
N0300 G99 G89 X-1.0 Y-4.0 R-6.75 Z-0.1 P100 F2.0	spotface hole 2
N0310 G98 X6.0	spotface hole 1, same depth as hole 2, then return to Initial for clearance
N0320 G99 Y7.5 R-6.75 Z-0.425	spotface hole 6 at ⅜-inch depth
N0330 X-2.0 L3	spotface holes 5, 4, and 3

The program can now be finished or continued as desired. Canned cycles may be initiated after a G80 or may just be changed from one canned cycle into the other "on the fly" without canceling in between. When a particular canned cycle is in effect, specific parameters may also be changed without requiring the reinitialization of the cycle. As in all other cases, thorough knowledge of the programming manual and the machine can extensively improve the efficiency of the program.

9.1.1.6 High-speed peck-drilling canned cycle G73

Also known as *woodpecker*, *intermittent feed*, or *deep hole drilling*, the high-speed peck-drilling canned cycle is used to break the drill swarf in holes whose depth exceeds 2.5 times their diameter. See Figure 9.8.

This canned cycle performs high-speed peck-drilling as follows: Rapid to *X-Y* location and to "*R*" level, feed by the amount "*Q*," rapid return by an unprogrammed fixed amount "*d*," feed and return alternately until "*Z*" is reached, and rapid return to "*R*" or Initial, depending on whether G99 or G98 has been programmed. The return amount "*d*" is usually set internally by control "parameters," and for our discussion it will be assumed to be 0.05. Our figure shows three "pecks" and two "interruptions" corresponding to three "*Q*"s and two "*d*"s, respectively. Since the total amount in feed, "*Z*" is shown as three times "*Q*" less twice "*d*," we can calculate "*Q*" by the following formula:

$$Q = \frac{\text{distance ``}Z\text{''} + (\text{Number of interruptions} \times \text{``}d\text{''})}{\text{Number of interruptions} + 1}$$

The distance "*Z*" represents the clearance above part + hole depth + 0.3 · drill diameter for clearance below the part. On some controls, one only needs to select

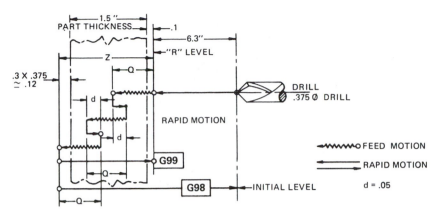

Figure 9.8   High-speed peck-drilling canned cycle G73.

a suitable "Q" value and the control will perform the necessary calculations and adjustments. In our figure,

$$Q = (1.72 + 2 \times 0.05)/3 = 0.61$$

and assuming the drill is in line with the required location, the tape block could read:

> N0100 G91 G98 G73 R-6.3 Z-1.72 Q0.61 F3.0

9.1.1.7 Peck-drilling canned cycle G83

The previous cycle was designed for self-removing swarf, which had to be broken ever so often. Some materials, however, generate chips that will stick to the drill flutes, pack them, and score the hole before breaking the drill. The peck-drilling canned cycle G83 will rapid to X-Y location and to "R" level, feed by the amount "Q," rapid return to "R," rapid in by "Q-d," feed "Q" from there, rapid return to "R," thus removing the swarf from the part, rapid in by "$2Q$-$2d$," feed "Q" from there, rapid return to "R," etc. It can be seen that the total traveling distance is higher than that of the high-speed peck-drilling cycle, and must therefore be used only when required by the material-tool-coolant combination. See Figure 9.9.

The tape block is identical to the previous example, with the exception of the code itself:

> N0100 G91 G98 G83 R-6.3 Z-1.72 Q0.61 F3.0

9.1.2 Turning centers

Advanced computer technology has allowed the CNC system builders to incorporate additional canned cycles in almost every new model.

In the past few years, these changes have increased the number of canned

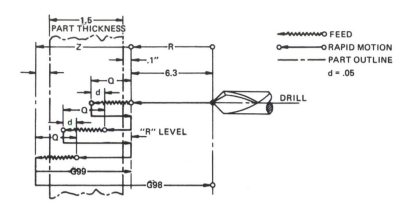

Figure 9.9   Peck-drilling canned cycle G83.

cycle options from a few to a practically unlimited number. As a result, it is almost impossible to describe the programming or the cycle patterns of all existing systems within the limitations of a single textbook. We shall, however, discuss the systems most commonly used in the industry today. Learning new cycles should not present difficulties to anyone who understands the workings of those described in the following pages.

Under the topic of machining cycles for turning centers, we will discuss *canned cycles* and *multiple repetitive cycles.*

Canned Cycles.   These cycles control four straight-line motions programmed in one tape block. The motions may be applied to turning, facing, or single-pass thread cutting. The first and last motions are in rapid, while the second and third are under controlled feed.

Multiple Repetitive Cycles.   These cycles, on the other hand, are used for the programming of stock removal in external and internal turning applications, peck-drilling, multiple groove turning, and multiple thread turning.

9.1.3 Canned Cycles

9.1.3.1 Straight-turning canned cycle G90

The straight-turning canned cycle can be used for a multitude of straight-turning or boring applications that require four straight-line motions. These four motions or cycle patterns are programmed in the same tape block, as opposed to four blocks in conventional programming.

The programming format for cylindrical external and internal turning is:

$$N\_\_G90X\_\_Z\_\_F\_\_$$

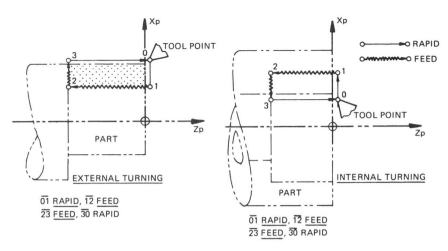

Figure 9.10   Straight-turning canned cycle G90.

The G90 code, not to be confused with absolute programming in machining centers, defines the cycle pattern of the tool point as illustrated in Figure 9.10. In both internal and external mode, we program the X- and Z-dimensions of point 2 after bringing the tool point to the position marked 0. To illustrate its use we will write a part program for the turning of a sample part illustrated by Figure 9.11.

N0010 G50 X0 Z0	Machine is at home position. Zero control
N0020 M41	Change gear to middle range
N0030 G98 T0101	Tool turret index to position 1, feed in ipm
N0040 S1200 M03	Spindle on, at 1,200 rpm
N0050 G00 X-8.0 Z-18.5	Rapid to point A
N0060 G50 X6.5 Z3.0	Transfer coordinate system from machine to part
N0070 G00 X3.85 Z.1 M08	Rapid tool-point to "0"
N0080 G90 X3.5 Z-1.5 F7.5	Turn 3.5-inch diameter to 1.5-inch length

The G90 canned cycle guides the tool point through $\overline{01}$ (rapid), $\overline{12}$ (turn outside diameter in feed), $\overline{23}$ (face to 1.5 in feed), and $\overline{30}$ (rapid). The programmed X- and Z-dimensions are to point 2 in the part coordinate system. Without the G90 code, the tool point would follow an entirely different pattern.

          N0090 X3.25 Z-1.0   Turn 3.25 diameter to 1.0 length

Since the G90 is modal we do not need to repeat it in subsequent tape blocks. The

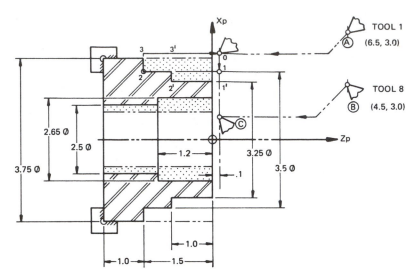

Figure 9.11   Sample part illustrating G90 canned cycle.

controlled motion pattern of block N0090 will be: $\overline{01'}$ (rapid), $\overline{1'2'}$ (feed), $\overline{2'3'}$ (feed) and $\overline{3'0}$ (rapid).

          N0100 G00 X6.5 Z3.0   Return tool to point A

The G00 has a dual function in the tape block N0100. In addition to placing the control into rapid mode, it will also *cancel* the *canned cycle*.

        N0110 G50 X-8.0 Z-1.85
        N0120 G30 U0.0 W0.0 T0808   Index tool turret to position 8
        N0130 X-10.7 Z-16.0 S1450   Increase speed to 1,450 rpm
        N0140 G50 X4.5 Z3.0   Reset the coordinate system to
                                        point B

The preceding block will establish the tool point's dimensional relationship to the part coordinate system.

        N0150 G00 X2.4 Z.1   Rapid tool point to "C"
        N0160 G90 X2.5 Z-2.55 F6.0   Turn 2.5-diameter-bore
        N0170 X2.65 Z-1.2   Turn 2.65-diameter bore to 1.20
                                        depth
        N0180 G00 X4.5 Z3.0   Rapid to point B, cancel G90
        N0190 G50 X-6. Z-18.0 M05   Transfer the coordinate system
                                        from part to machine

The programmed *X*- and *Z*-dimensions are not necessarily the actual distance between the tool point and the machine home. However, if in the next tape block, using the G28 code, we program *X*- and *Z*-values in between the current tool location (*B*) and the programmed home (in block N0190) position, the control will drive the turret slide back to the machine home position. If the machine is equipped with a tailstock, the programmer should ascertain that the return motion will take the turret around it.

N0200 G28 X-2.0 Z-6.M09   Return the tool turret to machine
home position

The above and other *return to home* blocks may be programmed only if at the beginning of the program (see block N0010) we have programmed a G50 X0 Z0 tape block to synchronize the control and the machine.

N0210 M30   End of program

(NOTE: If a tape block inside the canned cycle or following it contains only a sequence number and an end of block (EOB), the control will repeat the same cycle using the dimensions from the previous block.)
For example,

N0170 G90 X2.65 Z-1.2
N0180 E.O.B.

In N0180, N0170 will be repeated. *The G90 canned cycle must be canceled by a turret slide motion in G00 mode.*

9.1.3.2 Taper-turning cycle G90

The G90 canned cycle code can also be used to program tapered external and internal surfaces. Both the cycle pattern and the programming format must be changed to reflect the taper.
    The programming format for tapered external and internal turning is:

N__G90 X__Z__I__F__

The G90 controls the cycle pattern while the unit vector "*I*" defines the magnitude of the taper. The value of *I* may be positive (+) or negative (−), as illustrated in Figure 9.12.
    The unit vector "*I*" is programmed as the radius difference of the taper. The sign is viewed from the programmed point (2) to the feed start point (1). In either case, the clearance must be added to the part when we calculate the length of the tool pattern in feed.
    A typical industrial application is shown in Figure 9.13. To illustrate the use of the G90 code, we will write a part program for the turning of this part.

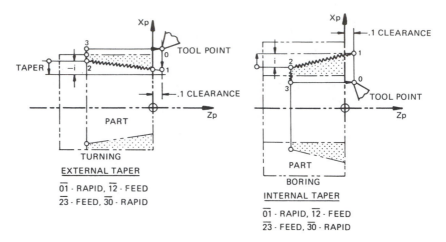

Figure 9.12   Taper turning canned cycle G90.

Figure 9.13 shows the taper as a difference of two diameters over a given length, to which we add the tool clearance. Therefore, the change in diameter on the 2-inch length will be 0.60 inch, or 0.30 inch in terms of the radius. This radius difference is then divided into as many passes (we selected three) as the job may require. Once the cycle pattern and dimensions have been established, the program can be written as follows:

```
N0010 G50 X0 Z0
N0020 M41
N0030 G98 T1111
N0040 S900 M03
N0050 G00 X-10. Z-18.5
```

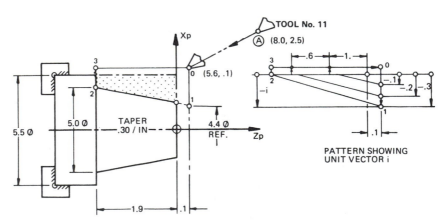

Figure 9.13   Sample part (tapered) illustrating G90 canned cycle.

Programming this rapid traverse to point "*A*" requires the programmer to check the clearance between turret and tailstock to avoid a tool collision. Alternatively, this motion block can be replaced by two tape blocks, first by programming the *Z*, and second by programming the *X*-motion.

N0060 G50 X8.0 Z2.5	Transfer coordinate system from machine to part
N0070 G00 X5.6 Z0.1	Rapid to start point "0"
N0080 G90 X5.0 Z-1.0 I-0.1	Canned cycle start, first taper turning pass
N0090 Z-1.6 I-0.2	Second pass in canned cycle
N0100 Z-1.9 I-0.3	Finish pass in canned cycle
N0110 G00 X8.0 Z2.5	Canned cycle cancel and rapid tool point to "A"
N0120 G50 X-10.0 Z-18.5 M09	Transfer coordinate system from part to machine
N0130 G28 X0 Z0 M05	
N0140 M30	

(NOTE: Internal tapers are programmed identically, except for the sign of the unit vector "I.")

9.1.3.3 Facing canned cycle G94

Facing canned cycle G94 is identical to the straight-turning cycle so far as the cycle pattern is concerned. The difference is that the cycle motion starts with the "*Z*" instead of the "*X*." The first and last (fourth) motions are rapid and the second and third motions are in feed. The programming format is:

$$N\_\_G94\ X\_\_Z\_\_F\_\_$$

The cycle can be applied to both internal and external facing operations. Typical patterns of this cycle are illustrated by Figure 9.14.

In both external and internal facing, we program the *X-Z* coordinates of point 2. The sample part illustrated in Figure 9.15 will outline the programming steps for the G94 code.

N0010 G50 X0 Z0	Zero control to machine
N0020 M41	Second gear
N0030 G40 T1100 M08	Cancel offsets, index tool and turn coolant on
N0040 G00 X-6.0 Z-18.5	Rapid to point "A"
N0050 G97 S500 M03	Constant rpm, spindle on
N0060 G50 X6.0 Z4.1	Transfer coordinate system from machine to part
N0070 G00 X6.5 Z0.1 T1111	Rapid to point "0", tool offset register 11

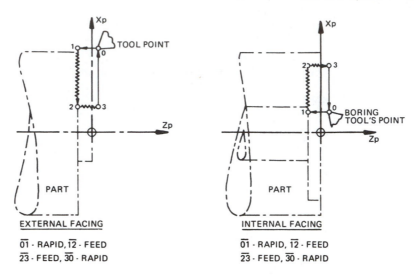

Figure 9.14   Facing canned cycle G94.

The value in tool offset register 11 is zero at start; the operator can adjust for any dimensional inaccuracy during the turning operation without changing the program.

N0080 G96 S450	Constant surface speed
N0090 G94 X1.0 Z-0.2 F.01	Turn 1.00 diameter, 2 length
N0100 X1.5 Z-0.4	Turn 1.50 diameter, 0.4 length
N0110 X2.0 Z-0.6	Turn 2.00 diameter, 0.6 length

The G96 in block N0080 will vary the spindle rpm to maintain the programmed 450 fpm cutting speed. The G96 should be programmed when the change in the

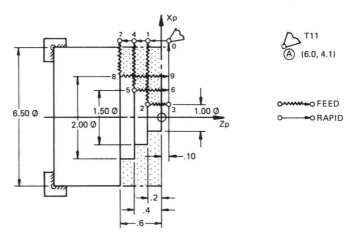

Figure 9.15   Sample part illustrating G94 canned cycle.

part diameter is larger than 1 inch. The motion pattern in blocks N0090, N0100, and N0110 can be described as:

N0090	$\overline{01}$—rapid, $\overline{12}$—feed, $\overline{23}$—feed and $\overline{30}$—rapid
N0100	$\overline{04}$—rapid, $\overline{45}$—feed, $\overline{56}$—feed and $\overline{60}$—rapid
N0110	$\overline{07}$—rapid, $\overline{78}$—feed, $\overline{89}$—feed and $\overline{90}$—rapid
N0120 G00 X8.2 Z4.1 M05	Rapid to point "A", cancel G94 canned cycle and stop spindle
N0130 G97 S500	Will change the constant surface speed to constant rpm
N0140 M30	End of program

This program has nine tape blocks less than an equivalent conventional part program without the G94 canned cycle. The program can be further improved by reducing the idle feed motion distances in blocks N0100 and N0110 from 0.5 inch ($\overline{56}$) and 0.7 inch ($\overline{89}$) to 0.3 inch and 0.5 inch, respectively. The change in the program can be implemented as follows:

N0090 G94 X1.0 Z-0.2 F.01	No change
N0100 G00 Z-0.1	Move canned cycle start point to Z-0.1-inch location
N0110 G94 X1.5 Z-0.4 F.01	Reprogram the canned cycle start point
N0120 G00 Z-0.3	Move canned cycle start point to Z-0.3-inch location
N0130 G94 X2.0 Z-0.6 F.01	Reprogram the canned cycle start point

The rest of the program remains unchanged. The saving in terms of machining time is insignificant in our example; however, it may be substantial in some other applications.

9.1.3.4 Taper face turning canned cycle G94

The programming is similar to the G90 tape turning canned cycle. The difference is the change of taper from the X- to the Z-axis, and the unit vector from I to K. The sign of the unit vector "K" is established with the same convention as we used for the cylindrical taper turning. The tape block format is written as:

N__G94 X__Z__K__F__

The cycle patterns for the G94 canned cycle are illustrated in Figure 9.16 for both internal and external face turning.

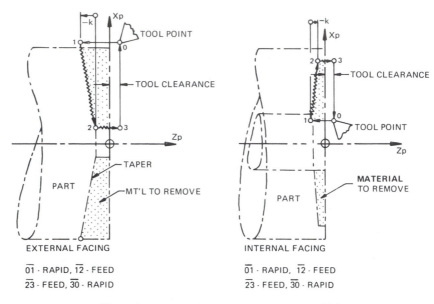

Figure 9.16 Taper-face turning canned cycle G94.

The *X*- and *Z*-dimensions of point 2 are programmed in the canned cycle tape block. The direction of the unit vector "*K*" is viewed from this point to the start of the feed motion (point 1). If this direction (indicated by the arrows) is that of the (+) positive "*Z*-" motion, the sign of "*K*" will also be positive (+). Alternatively, the sign of "*K*" will be negative.

We will use the sample part shown in Figure 9.17 to outline the programming steps for the taper face turning canned cycle, G94.

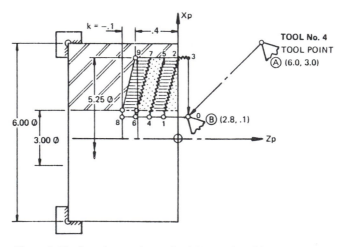

Figure 9.17 Sample part (taper face) illustrating G94 canned cycle.

If the taper dimensions are readily available from the part drawing, as is the case in our example, we can proceed with the programming. However, if the taper is specified as: "Taper per inch of diameter" (0.02/1-inch-diameter), the programmer must first convert this information into programmable dimensions. The calculation in most cases will consist of simple additions or subtractions. This part program can now be written as:

```
N0010 G50 X0 Z0
N0020 M41 G99
N0030 G40 T0400 M08          Cancel offset, index tool, coolant on
N0040 G00 X-5.5 Z-16.5       Rapid to point "A"
N0050 G97 S900 M03           Constant rpm, spindle on
N0060 G50 X6.0 Z3.0          Transfer coordinate system from machine to
                                part
N0070 G00 X2.8 Z0.1 T0404    Rapid to start point "B"; at start, register 04 is
                                set to zero. Subsequently, the operator can
                                adjust for dimensional inaccuracies without
                                having to change the program.
N0080 G96 S380               Constant surface speed
N0090 G94 X5.249 Z0 K-0.1 F .01   First pass in G94
```

We have programmed 5.249 inches, instead of the part dimension of 5.25 inches. If in each subsequent tape block, we further reduce the diameter by the same amount, then, after the fourth pass we can program a finishing cut. The few thousands left on for the finish cut will reduce the cutting forces substantially, resulting in better accuracy and surface finish.

```
N0100 X5.248 Z-0.2       Second pass
N0110 X5.247 Z-0.3       Third pass
N0120 X5.246 Z-0.395     Final roughing pass
N0130 X5.25 Z-0.4        Finishing pass
```

This method of rough and finish turning with the same tool is most economical for small (under 50 parts) production runs. If the finish turning is done with another tool (in the same setup), the program should be written to leave 0.010 to 0.050 inch on diameters and 0.005 to 0.025 inch on faces.

```
N0140 G00 X6.0 Z3.0 M09    Cancel G94 canned cycle and rapid
                              to point "A"
N0150 G97 S900             Constant rpm
N0160                      End of program
```

Once the unit vector is programmed, as shown in tape block N0090, it need not be repeated in subsequent blocks.

(NOTE: TNR compensation can be programmed for both G90 and G94 canned cycles. It should be programmed in the block preceding the G90 or G94 canned cycle.

9.1.4 Multiple Repetitive Cycles

These are more complex than canned cycles. It is in this area that the CNC systems vary the most. These features are developed for special applications and involve substantial software programming by the manufacturer of the control unit.

9.1.4.1 Stock removal cycle G71

Stock removal cycle G71 is used when several turning passes are required to remove stock. The result is a part outline bounded by straight lines, slopes, and circular arcs. The tape format is developed to adapt to any shape or form of turning or boring cycle pattern.

The tape format below applies to General Numerics controls. Other controls may use slightly different formats.

$$N\_G71\ P(ns)\ Q(nf)\ U(\Delta u)\ W(\Delta w)\ D(\Delta d)\ F\_S\_$$

The cycle pattern for this tape block is illustrated in Figure 9.18.

P(ns)   "P" is the address of the tape block number where the cycle starts (ns). This number is normally the number following the tape block with the G71 code.

Q(fs)   "Q" is the address of the tape block number where the cycle ends (fs).

U(Δu)   "U" is the address of the material's dimension (Δu) left for finish turning on diameters.

W(Δw)   "W" is the address of the material's dimension (Δw) left for finish turning on faces.

D(Δd)   "D" is the address for the depth (Δd) cut. The Δd dimension cannot be programmed with decimal point. Example: 0.2-inch depth is D2000, 0.15-inch depth is D1500, 0.06-inch depth is D600

"F" and "S" are feed and speed word addresses.

The format within the cycle is restricted. Codes such as *S*, *T*, *M*, or G96, G97, may not be programmed between the starting sequence number, defined by

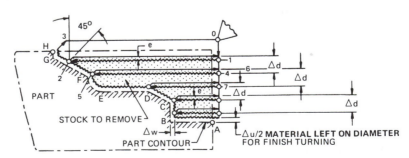

Figure 9.18   Cycle pattern for stock removal cycle G71.

"*P*" and the ending sequence number defined by "*Q*." These codes, if required, must be programmed prior to the G71 cycle block. The control will maintain these programmed codes for the validity of the multiple repetitive cycle. As shown in Figure 9.18, "*e*" is the amount by which the tool point will retract from the part surface. This motion occurs at the end of each feed pass on a 45° angle. The cycle tape block sets up a subprogram to calculate the number of cuts or passes our tool must complete to produce the part contour. This contour is programmed in the tape blocks following the G71 code. Each segment or element is programmed by conventional means, using G01, G02, or G03, as the case may be, from point "*A*" to point "*H*." The cycle must be terminated by a G00, rapid motion block. The G71 code can be used in the same manner for internal contour turning.

For the pattern shown in Figure 9.18, the tool point path can be described as follows:

First pass: Rapid to point 1, feed to point 2, feed to point 3 following the part contour, rapid by "e" at 45° toward start point, rapid return.

Second pass: Rapid to point 4, feed to point 5, feed to point 2 following the part contour and leaving the programmed (Δu) or (Δw) stock on for finish turning; rapid by "e" at 45° and rapid return.

The control will guide the tool point through the remaining passes, sometimes called "loops," until the contour is completed. Should the last loop be smaller than the Δd dimension, the control will automatically adjust the dimension.

To demonstrate the use of the G71 code, we will write a part program for the turning of a part illustrated in Figure 9.19. The function of each block and code will be explained in detail.

N0010 G50 X0Z0	Zero control to machine
N0020 G99 M41	Inches per revolution—ipr
N0030 G00 G40 T0300	Tool offset cancel, tool change to position 3
N0040 G00 X-6.8 Z-16.0	Rapid to point "A"
N0050 G50 X8.5 Z2.5 S1200	Transfer coordinate system from machine to part, set max. rpm at 1,200
N0060 G97 S800 M03	Start spindle at 800 rpm
N0070 G00 X7.6 Z0.1	Rapid to point "B"
N0080 G96 S300 M08	Constant surface speed; as the diameter of part decreases, the speed will be increased to maintain 300 fpm cutting speed
N0090 G71 P0100 Q0200 U0.04 W0.02 D2000 F0.012	This block initiates the stock removing cycle; the subprogram which describes the part contour starts in block 0100 (P0100) and ends in block 0200 (Q0200); 0.04/2 = 0.02 inch will be left

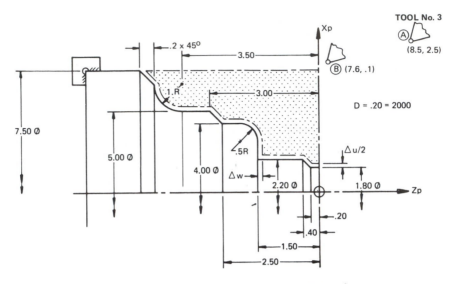

Figure 9.19   Sample part illustrating the G71 cycle.

on all the diameters (U0.04) and 0.02
inch on all the faces (W0.02). Depth of
cut is set at 0.2 inch (D2000).

```
N0100 G00 X1.8
N0110 G01 Z-0.2 F0.01
N0120 G01 X2.22 Z-0.4
N0130 Z-1.5
N0140 X3.0
N0150 G03 X4.0 Z-2.0 R0.5
N0160 G01 Z-2.5
N0170 X5.0 Z-3.0
N0180 Z-3.5
N0190 G02 X7.0 Z-4.5 R1.0
N0200 G01 X7.5 Z-4.7
```

From sequence N0100 to sequence N0200 the program describes the part contour.
Even though the first block N0100 defines a tool position of $X = 1.80$ diameter,
the tool point will start the cycle at 0.2 inch (D2000) depth of cut. It will, however,
define a part boundary into which the cutter or tool point will not enter. The "*P*"
block (N0100) must only contain an *X*-motion.

N0210 G00 X8.5 Z2.5	Cycle cancel and return the tool point to point "A"
N0220 G97 S800 M09	Constant rpm

N0230 G50 X-6.8 Z-16.0	Transfer coordinate system from part to machine
N0240 G28 X0.0 Z0.0	Return turret slide to machine home position
N0250 M30	End of program

Internal contour turning can be programmed using the same format and programming steps.

9.1.4.2 Stock removal in facing G72

Stock removal in facing G72 is used when the repetitive turning motions are parallel with the X-axis. Accordingly, the "P" block must only contain a Z-motion. Examples of internal and external face turning using the G72 cycle are illustrated in Figure 9.20.

The programming format is identical to the turning and boring cycle discussed previously, except for the depth of cut (d), which switches from the X- to the Z-direction.

The tool point path can be described as follows: Rapid to point 1 by the amount Δd, straight-line feed parallel to the X-axis, followed by feed along the part contour, rapid by amount "e" at 45°, and rapid return to point 0.

The magnitude of the motion "e" is set by internal parameter. This cycle is then repeated by the control until the final depth is equal to or smaller than the "d" distance. After the final cut, the tool point will return to the start point 0. It should be noted that the tool point motion in feed in this particular feed cycle must either steadily increase in size or decrease from the start to the end. The control cannot perform this cycle if this requirement is not satisfied due to a profile containing a reverse contour.

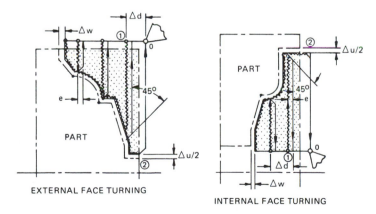

EXTERNAL FACE TURNING   INTERNAL FACE TURNING

Figure 9.20   Cycle patterns for the G72 code.

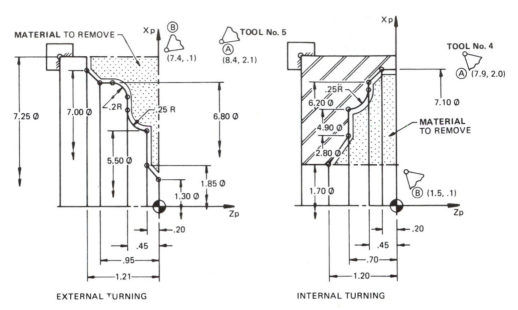

Figure 9.21 Sample part illustrating the application of G72.

The next part program will demonstrate the use of the G72 code. Figure 9.21 shows only the portion of the part directly related to the program.

External Turning	
N0010 G50 X0.0 Z0.0	Zero control to machine
N0020 G99 M41	ipr, second gear of headstock
N0030 G00 G40 T0500	Tool offset cancel, tool change to turret position 5
N0040 G00 X-4.5 Z-20.	Rapid to point "A"
N0050 G50 X8.4 Z2.1 S990	Transfer coordinate system from machine to part and set maximum speed to 990 rpm
N0060 G97 S350 M03	Start spindle at 350 rpm
N0070 G00 X7.4 Z.1	Rapid to start point "B"
N0080 G96 S400 M08	Constant surface speed

The change in diameter is 5.95 inches; therefore, it is necessary that a constant surface speed block be programmed. The constant surface speed not only will improve the program efficiency, but it will also maintain a constant cutting condition (load on tool, chip breaking, etc.).

N0090 G72 P0100 Q 0170 U.05 W.02 D1500 F.01

We shall now initiate the stock removing cycle G72 and set the limits for rough turning of the contour, as programmed in the following tape blocks:

```
N0100 G00 Z-1.21
N0110 G01 X 7.0 F0.007
N0120 X 6.8 Z-0.95
N0130 Z-0.65
N0140 G02 X6.4 Z-0.45 R0.2
N0150 G01 X 6.0
N0160 G03 X 5.5 Z-0.2 R0.25
N0170 G01 X1.85
N0180 X 1.3 Z0
```

Between sequences 0100 and 0180, the program describes the part boundary. The tool point will stay out of this boundary by the set $U/2$ and W dimensions programmed in block N0090. The first motion block inside the cycle or loop contains a Z-motion only.

N0190 G00 X 8.4 Z2.1	cancel G72 cycle and return the tool point to point "A" in rapid traverse
N0200 G97 S350 M09	change from constant surface speed to constant rpm
N0210 G50 X-4.5 Z-20.0	transfer coordinate system from part to machine
N0220 G28 X0.0Z0.0 M09	return turret slide to machine home position and turn coolant off
N0230 M30	end of program

You know, at this stage, how the G72 code controls the tool point during the facing cycle. The following Internal Turning program will not contain additional explanations.

```
Internal Turning
N0010 G50 X 0.0 Z0.0
N0020 G99 M41
N0030 G00 G40 T0400
N0040 G00 X-5.5 Z-16.5
N0050 G50 X 7.9 Z2.0 S1000
N0060 G97 S950 M03
N0070 G00 X1.5 Z0.1 M08
N0080 G96 S375
N0090 G72 P0100 Q0160 U-0.04 W0.015 D1000 F0.012
N0100 G00 Z-1.2909
N0110 G01 X2.8 Z-0.7 F0.006
N0120 X4.9
N0130 G03 X5.4 Z-0.45 R0.25
```

```
N0140 G01 X6.2
N0150 X7.1 Z-0.2
N0160 Z0.0
N0170 G00 X7.9 Z 2.0 M09
N0180 G97 S950
N0190 G50 X-5.5 Z-16.5
N0200 G28 X0.0 Z0.0 M05
N0210 M30
```

The preceding programming examples should provide sufficient exposure into repetitive cycles for rough turning. The basic principles and steps for other controls are either identical or quite similar. It is recommended that you study the tape format and cycle patterns provided in the system programming manual, prior to attempting to write a part program.

9.1.4.3 Finish turning cycle G70

The finish turning cycle G70 is also known as single-pass contouring cycle. It can only be programmed after a G71 and/or a G72 multiple turning cycle within the same part program. As discussed earlier, the finish turning can be done using the same tool as for roughing in small production runs. However, a different tool should be used for medium-to-large production runs. The G70, like the G71 and G72 cycles, can only be performed in memory mode of operation. The tape format is as follows:

$$N\_G70\ P(ns)\ Q(nf)$$

The P and Q words must be identical to the ones programmed for rough turning. The finish turning tape block does not allow programming of T, S, or F commands; therefore, the tool code and speed codes must be programmed after the end of the roughing and before the finishing tape block. The feed code, on the other hand, should be programmed inside the loop of the rough turning cycle. This feed will be used only during the finishing cycle. Should the programmed feed be too coarse for the finishing, the programmer can compensate by using a higher speed, which can be programmed prior to the finishing cycle block (feed in ipm).

To demonstrate the use of the G70 code, we will alter the programs written for the rough turning of the parts shown in Figure 9.19 and 9.21. Instead of rewriting all the programs, we shall insert additional tape blocks as the application may require.

Finish turning of the part shown in Figure 9.19. Tape blocks N0010 to N0220 require no changes.

N0230 M01   Optional program stop in the "on" position will allow the operator
                                                        to inspect the part; the operator can override the M01 in the
                                                        "off" position

N0240 G97 S1100 M03	Start spindle at 1,100 rpm. When the M01 switch is turned off, the spindle speed will be increased to 1,100 rpm
N0250 G00 X7.6 Z0.1 M08	Rapid tool point to start position, coolant on
N0260 G96 S500	Constant surface speed has been increased from S300 (N0080) to S500. The increased surface speed will result in a higher (66%) rpm and higher quality surface finish even though the programmed feed rate of F0.01 has not been changed.
N0270 G70 P0100 Q0200	Initiates the finish turning cycle. The tool point will be guided along the part boundary derived from Seq. No. N0100 to N0200 inclusive
N0280 G97 S800	Constant rpm
N0290 G00 X8.5 Z2.5	Rapid the tool point back to point "A"
N0300 G50 X-6.8 Z-19.5 M09	Transfer coordinate system from part to machine
N0310 G28 X0.0 Z0.0	Return turret slide to machine home position
N0320 M30	End of program

Since the finish turning cycle is utilizing the tape block written for rough turning, it is mandatory that the first tape block (N0100) inside the loop be programmed in rapid motion. The second block (N0110) is for machining; therefore, it must be in linear (G01) or circular (G02 or G03) interpolation mode, combined with the appropriate feed code.

Finish turning of the parts shown in Figure 9.21: external turning. Tape blocks N0010 to N0180 will not change.

```
N0190 M01
N0200 G97 S500 M03
N0210 G00 X7.4 Z0.1 M08
N0220 G96 S600
N0230 G70 P0100 Q0180
N0240 G97 S350 M09
N0250 G00 X8.4 Z2.1 M05
N0260 G50 X-4.5 Z-20.
N0270 G28 X0.0 Z0.0
N0280 M30
```

Internal turning. Tape blocks N0010 to N0160 will not change.

```
N0170 M01
N0180 G97 S1150 M03
N0190 G00 X7.9 Z2.0 M08
N0200 G96 S525
N0210 G70 P0100 Q0160
N0220 G97 S950 M09
N0230 G00 X7.0 Z2.0 M05
```

```
N0240 G50 X-5.5 Z-16.5
N0250 G28 X0.0 Z0.0
N0260 M30
```

9.1.4.4 Pecking cycle G74

The pecking cycle G74, a multiple repetitive cycle pattern, is identical to the peck-drilling canned cycle discussed earlier. The repetitive motion of the tool turret can be used for either turning grooves in the part face or drilling a hole in the center of the part. In both cases, the cutting process requires an interruption of the feed motion to break up the continuity of the chip. The pecking-cycle standard on most turning centers for drilling is an optional feature for grooving.

9.1.4.5 Multiple grooving cycle G74

This cycle is used for the turning of deep multiple grooves. Since the turning cycle controls the Z-axis in feed and the X-axis in rapid motion, we can turn any number of grooves using only one tape block.

If we assume a feed motion of $K = 0.2$ inch, it would take a minimum of 18-tape blocks in conventional programming. Instead, we can write one single tape block as follows:

$$N\_\_G74\ X\_\_Z\_\_I\_\_K\_\_F\_\_$$

using the G74 cycle.

The individual addresses are marked on the cycle pattern drawing shown in Figure 9.22.

This cycle allows turning equally spaced grooves. The "*e*" rapid retract dimension is set by an internal parameter.

To demonstrate the use of the G74 peck-grooving cycle, we shall write a part

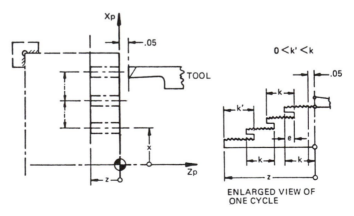

Figure 9.22 Cycle pattern for pack-grooving cycle G74.

program for the turning of the sample part illustrated in Figure 9.23. The dimensions for each address of the G74 cycle are established as follows:

X = 3.650 diameter, diameter of the last groove turned in the cycle. The turning sequence may be reversed to start the groove at 3.65 diameter; however, the X-dimension would have to be changed to 5.47 diameter. This value was calculated as follows:

$$3.65 + 2(0.455 + 0.455) = 5.47$$

Z = −0.6, the depth of the groove

I = 0.455, distance between the grooves; this distance cannot vary from groove to groove.

K = 0.2, distance of Z-motion in feed between interruptions. K is selected by the programmer. Both I and K should be programmed unsigned.

F = 0.006, turret slide velocity during the feed motion. F is also selected by the programmer.

Once these address values are established, we can write the part program by using the steps of sequence outlined below:

```
N0010 G50 X0.0 Z0.0
N0020 G99 M41
N0030 G00 G40 T0300
N0040 G00 X-8.5 Z-18.5                Rapid tool to point "A"
N0050 G50 X6.5 Z2.5 S1200            Transfer coordinate system from machine to part and
                                       set maximum spindle rpm

N0060 G97 S650 M03
N0070 G00 X5.47 Z0.05 M08            Rapid tool point to point "A"
N0080 G96 S450                      Set constant surface speed at 450 fpm
N0090 G74 X3.65 Z-0.6 I0.455 K0.2 F0.006
N0100 G00 X6.5 Z2.5 M09
N0110 G97 S650
N0120 G50 X-8.5 Z-18.5 M05
N0130 G28 X0.0 Z0.0
N0140 M30                           End of program
```

This short program could easily be changed to turn any number of grooves by changing the dimensions of the "X" and "I" addresses in the G74 cycle block.

9.1.4.6 Peck-drilling cycle G74

The peck-drilling cycle G74 is used for drilling deep holes. The tape format is different from the peck grooving. Since the hole can only be drilled in the center of the part, there is no X-motion in the cycle. The subroutine, albeit using the same G74 code, can distinguish between the drilling and the turning cycles because of the change in the format.

N__G74 Z__K__F__

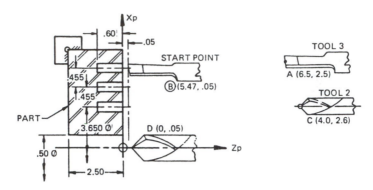

Figure 9.23 Sample part illustrating the G74 cycle.

We shall write a part program for the drilling of the 0.50-diameter hole in the center of the part shown in Figure 9.23.

```
N0010 G50 X0.0 Z0.0
N0020 G99 M41
N0030 G00 G40 T0202
N0040 G00 X-8.2 Z-16.
N0050 G50 X4.0 Z2.6 M08
N0060 S850 M03
N0070 G00 X0.0 Z.05
N0080 G74 Z-2.65 K0.5 F3.5
```

The G74 cycle will start the peck drilling to a total Z-depth of -2.65, including the tool point angle. The cycle will repeat the interruption at every 0.5 ($K = 0.5$) distance until the total depth is drilled, then the tool point will return in rapid motion to the starting point "D."

```
N0090 G00 X4.0 Z2.6 M09
N0100 G50 X-8.2 Z-16.0 M05
N0110 M30                        End of program
```

This cycle bears no resemblance to the peck-drilling cycles discussed under the machining centers. This cycle was developed strictly for the drilling of one single hole per part.

9.1.4.7 Multiple groove turning cycle G75 (In the *X*-axis Direction)

The multiple groove turning cycle G75 pattern is identical to G74 if the X- and Z-axes, as well as the "*I*" and "*K*" addresses, are reversed. The two programming formats will therefore be very similar. The function of the individual addresses can be described as follows:

X = bottom diameter of the grooves; this must be the same for all the grooves
turned in the cycle block
Z = dimension from Z-part zero to the last groove in the cycle
I = the depth of the tool point motion in feed between interruptions
K = the dimension between the equally spaced grooves in the cycle

The use of the G75 groove turning cycle will be shown in Figure 9.24.

The program may be written with the starting groove nearest to the chuck
or the tailstock by changing the Z-dimension in the block preceding the G75 cycle
block.

N0010 G50 X0.0 Z0.0
N0020 G99 M41
N0030 G00 G40 T0100
N0040 G00 X-6.0 Z-18.5
N0050 G50 X6.5 Z1.75 S1200
N0060 G97 S850 M03
N0070 G00 X3.45 Z-0.5 M08

N0080 G96 S400   The constant surface speed will produce constant cutting
                                      conditions. As a result, even surface finish.

N0090 G75 X2.25 Z-2.0 I.2 K.5 F.005   As in the previous example, the "I" and "K" values are
                                      programmed unsigned.

N0100 G00 X6.5 Z1.75
N0110 G50 X-6.0 Z-18.5
N0120 G97 S850 M09
N0130 G28 X0.0 Z0.0 M05
N0140 M30   End of program

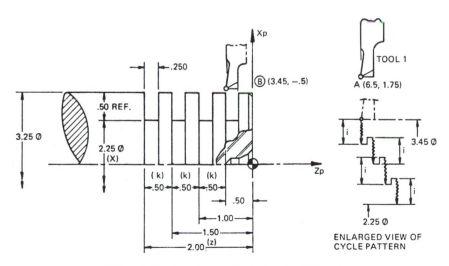

Figure 9.24   Sample part illustrating the G75 cycle.

9.1.5 Thread Cutting

CNC system builders have developed specific canned cycles for the turning of straight, tapered, and scroll threads for single, double, and multiple starting threads. Most current CNC turning centers are equipped with straight- and taper-threading cycles for both internal and external thread turning as standard features. More complex cycles are offered on an optional basis at extra cost. In the following pages, we shall attempt to clarify in some detail the cycle patterns, tape formats, and the programming of thread turning. Because of their similarities, the threading cycles are presented as a separate group.

9.1.5.1 Single start multipass-thread-turning cycle G92

The single start multipass-thread-turning cycle G92 is the simplest canned cycle for the turning of straight and tapered threads for internal and external applications. The cycle pattern is identical to the G90 turning and boring cycle from which it was derived. While the speed-feed ratio had no significance during the G90 cycle, in *threading, the speed-feed ratio has prime importance.* As we have earlier discussed, the lead of the thread is a function of the spindle rpm. If the tool point moves at 1 ipm, and the spindle is rotating at 100 rpm, the lead (pitch) of the thread will be 0.01 ipr. To produce a constant lead from the beginning to the end of the thread, the spindle must maintain a constant rpm. For this reason, all *the thread-turning cycles lock the programmed spindle speed and turret slide feed* during the cycle. Any adjustment of the speed or feed override circuits is ignored by the control. The programmer is advised to study the appropriate section of the programmer's manual for restrictions such as maximum rpm, minimum and maximum programmable lead, etc. These programmable parameters vary from system to system even for the same make of machine and control.

The 4NE turning center illustrated in the following discussion allows the programmer to use conventional feed programming to an accuracy of four decimal places. In this mode, the range of the feed address can vary from a minimum of F0.0001 to a maximum of F50.0000 ipm. Turning a part with 18 tpi would require $\frac{1}{18} = 0.055556$ ipr feed rate. Since this feed programming format will allow us to program F0.0555 feedrate only, for longer threads, the inaccuracy can add up to several thousands of an inch. To compensate for this inaccuracy, we have the option of using an "E" instead of an "F" address. In the "E" mode, we can program our feed to six-decimal-point accuracy. The range of the "E" address can vary from E0.000001 to E9.999999 ipr, providing a more flexible method for turning threads to gage accuracies.

The *maximum* programmable spindle speed (n) must be smaller or equal to a ratio of 196.85 (constant) divided by the lead (L), expressed in mathematical terms as:

$$n = \frac{196.85}{L}$$

For 12 tpi, the lead (pitch) $L = \frac{1}{12} = 0.0833333$, and the maximum rpm ($n$) must be less than or equal to:

$$n = \frac{196.85}{0.0833333} = 2,362 \text{ rpm}$$

Obviously this is far too high. Taking another example of 6 tpi, twice the 12 tpi:

$$n = \frac{196.85}{0.166666} = 1,181 \text{ rpm}$$

This is still a very high rpm for thread turning. See Figure 9.25.

The programming format is as follows:

$$N\_\_G92\ X\_\_Z\_\_F\ (or\ E)\_\_$$

where

N is the sequence number

X is the tool point dimension at start point

F is the feed in ipr to four decimal places accuracy or

E is the feed in ipr to six decimal places accuracy.

Only one "F" or "E," but not both, can be used in the cycle block.

The cycle patterns for the G92 straight-thread-turning cycle are shown in Figure 9.25.

The tool point must be at point 0 and the spindle in rotation prior to pro-

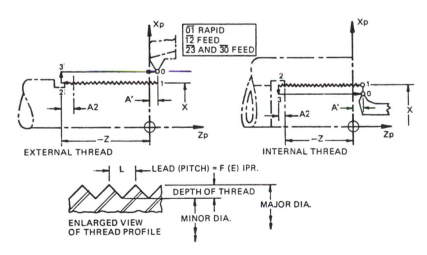

Figure 9.25   Cycle pattern for the straight-thread-turning cycle of G92.

gramming the G92 thread-turning cycle. Before discussing a sample program, we must look at the effect of the tool's acceleration A1 and deceleration A2 at the beginning and end of the thread. The dimension of A1 is the distance required by the tool point to reach the programmed feed rate. A2 is the distance required by the tool point to come to a stop from the programmed feed rate. These dimensions can either be calculated by a formula, provided in the programming manual, or taken from a chart or graph. For the 4NE turning center, the A1 and A2 dimensions can be calculated using the formulas below:

Spindle rpm	A1	A2
100	0.41 × Lead	0.2 × Lead
200	0.82 × L	0.4 × L
300	1.23 × L	0.6 × L
400	1.64 × L	0.8 × L
500	2.05 × L	1.00 × L
600	2.46 × L	1.15 × L
700	2.86 × L	1.4 × L
800	3.27 × L	1.6 × L
900	3.68 × L	1.8 × L
1,000	4.09 × L	2.0 × L
1,100	4.50 × L	2.2 × L
1,200	4.91 × L	2.4 × L

Turning an 8 tpi part at 100 rpm would require an acceleration distance of:

$$L = \frac{1}{8} = 0.125$$

$$A1 = 0.41 \times 0.125 = 0.0512 \text{ inch}$$

and a deceleration distance of:

$$A2 = 0.2 \times 0.125 = 0.0250 \text{ inch}$$

However, turning the same thread at 1,200 rpm would require distances of:

$$A1 = 4.91 \times 0.125 = 0.6137 \text{ inch}$$

and

$$A2 = 2.4 \times 0.125 = 0.3 \text{ inch}$$

The programmer must therefore select the rpm to suit the clearances available on the part. The range is normally adequate to turn the thread to the designed tolerances.

9.1.5.2 Tapered thread-turning using G92

The programmer must know the maximum angle (α) of the taper which can be turned on a particular CNC lathe. For the 4NE turning center, the maximum angle (α) is 45°.

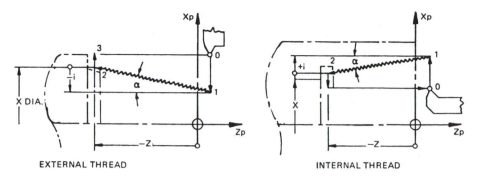

EXTERNAL THREAD   INTERNAL THREAD

Figure 9.26   Cycle pattern for the taper-thread-turning cycle G92.

The tape format for the programming of the G92 code for tapered thread is as follows:

$$N\_G92 \ X\_Z\_I\_F(or \ E)\_$$

The cycle pattern is shown in Figure 9.26. To demonstrate the use of the G92 code, we will write a part program for threads turning on the part, shown in Figure 9.27.

It will be assumed that the turning portion has already been completed. The following steps are required:

1. Calculate the lead $F(E)$ for both tools:
   Tool No. 2   8 tpi = $\frac{1}{8}$ = 0.125
   Tool No. 7 12 tpi = $\frac{1}{12}$ = 0.803333
2. Calculate the distances A1 and A2 required for the tool point acceleration and deceleration:

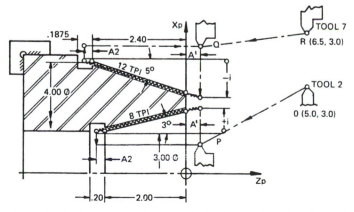

Figure 9.27   Sample part illustrating the G92 taper-thread-turning cycle.

Selecting 400 rpm for tool No. 2 and 300 rpm for tool No. 7, the distances required can be worked out.

$$\text{Tool No. 2—A1} = 1.64 \times 0.125 = 0.205; \text{round off to } 0.21$$
$$\text{A2} = 0.8 \times 0.125 = 0.1 \text{ inch}$$
$$\text{Tool No. 7—A1} = 1.23 \times 0.083333 = 0.1025; \text{round off to } 0.11$$
$$\text{A2} = 0.6 \times 0.083333 = 0.05$$

At these selected spindle speeds, both tools will have enough clearance to decelerate, therefore, our values will not affect the accuracy of the threads.

3. Establish the start point dimensions for tool No. 7: Q (4.5, 0.11); for tool No. 2: P (2.5, 0.21). These values will allow adequate clearances at the end of the feed stroke for both tools to clear the part.

Having established the above three sets of dimensions, we can now proceed with the part program

```
N0010 G50 X0 Z0
N0020 M40 T0202 G99                    tool No. 2
N0030 G97 S400 M03
N0040 G00 X-5.4 Z-15.
N0050 G50 X5.0 Z3.0 M08
N0060 G00 X2.5 Z0.21                   rapid to point "P"
N0070 G92 X2.8728 Z-2.1 I0.1208 F0.125  first pass
```

$$Z = -2.0 + (-A2) = -2.0 - 0.1 = -2.1$$
$$I = (2 + 0.1 + 0.205) \cdot \tan 3° = 0.1208$$

The G92 has set the cycle pattern for the multipass thread turning. The thread turned by tool No. 2 is a class 3 thread. Its major diameter is 3.000 inches, and its minor diameter is 2.8723 inches. From practical experience, six passes are required. As the depth increases, we will progressively reduce the depth of cut. The final pass for finishing will be repeated at the same depth to size the thread; accordingly, the cutting force in the finish pass will be approximately 5% of the preceding cutting pass, without any tool deflection. This will produce a part with uniform surface finish and accurate tolerance.

```
N0080 X2.9128    second pass
N0090 X2.9428    third pass
N0100 X2.9728    fourth pass
N0110 X2.9895    fifth pass, finished minor diameter
N0120 X2.9895    final pass
```

The programmed values of the *Z*, *I*, and *F* addresses will remain active and be repeated in each pass from N0080 to N0120.

N0130 G00 X5. Z3. T0200	cancel G92 cycle
N0140 S300 T0707	tool No. 7
N0150 G50 X6.5 Z3.0	
N0160 G00 X4.5 Z0.11	rapid to Point "Q"
N0170 G92 X3.9687 Z-2.45 I-0.2233 E0.083333	first pass

The major diameter of the 4-12 tpi thread = 4.00, and the minor diameter = 3.8917. We will use again six passes for the thread turning.

$$Z = -2.4 + (-A2) + (-A1) = -2.4 - 0.05 - 0.1025 = -2.5525$$

$$I = 2.5525 \cdot \tan 5° = -0.2233$$

For accuracy, we have replaced the *F* address by *E*.

N0180 X3.9387	second pass
N0190 X3.9187	third pass
N0200 X3.9087	fourth pass
N0210 X3.9009	fifth pass, finished minor diameter
N0220 X3.9009	final pass
N0230 G00 X6.5 Z3.0 M09 T0700	cancel G92 cycle
N0240 G50 X-5.4 Z-15.0 M05	
N0250 G28 X0 Z0	
N0260 M30	

The latest CNC turning centers now offer an even more compact thread-turning cycle. This cycle allows the programmer to specify all the data required for any thread turning in one tape block.

9.1.5.3 Multipass thread-turning cycle G76

The multipass thread-turning cycle G76 allows the programmer to define any number of passes in one tape block. The CNC system will perform straight and tapered and internal and external threading. The guidelines for the spindle rpm and tpe calculations, outlined for the G92 cycle, do not change. The tape format is as follows:

N_G76 X_Z_I_K_D_ F(E)_A_

where G76 is the multipass thread cycle,

X   = the minor diameter of the thread.
Z   = the length of thread plus the deceleration distance.

I = the radius difference measured between the start and end points of the
        thread

K = the depth of the thread, calculated as $K = \dfrac{D \text{ major} - D \text{ minor}}{2}$

D = the depth of cut for the first pass (measured in terms of the radius, no
        decimal point)

$F(E)$ = the lead of the thread

A = the angle of thread, normally 30°, 55° or 60°. The 4NE has additional 0°,
        29°, and 80° programmable angles.

Since the cycle pattern is identical to the G92 code, we will not repeat the
explanations. Instead, we shall write a part program for the thread turning of the
same part. Using the same thread and tool data, we can proceed with the part
program as:

```
N0010 G50 X0 Z0
N0020 M40 T0202
N0030 G97 S400 M03
N0040 G00 X-5.4 Z-15.0
N0050 G50 X5.0 Z3.0 M08
N0060 G00 X2.5 Z0.21
```

These tape blocks are identical to our previous program.

```
N0070 G76 X2.9895 Z-2.1 I0.1208 K0.0811 D150 F0.125 A60
```

This G76 cycle block replaced the tape blocks N0070 to N0120 and will
produce the same 3″-8 tpi thread. The dimensions used are identical but programmed in different formats. "K" is calculated as follows:

$$K = \frac{2.9895 - 2.8273}{2} = 0.0811$$

The thread profile angle "A" is defined by the part drawing.

```
N0080 G00 X5.0 Z3.0 T0200
N0090 S300 T0707
N0100 G50 X6.5 Z3.0
N0110 G00 X4.5 Z0.11
N0120 G76 X3.9009 Z-2.45 I-0.2233 K0.0540 D100 E0.083333 A60
```

The value of "K" is derived as follows:

$$K = \frac{4.0088 - 3.9009}{2} = 0.0540$$

N0130 G00 X6.5 Z3.0 T0700 M09
N0140 G50 X-5.4 Z-16.5 M05
N0150 G28 X0.0 Z0.0
N0160 M30   end of program

We have reduced the program from 26 to 16 tape blocks. It would also require proportionally less memory core for the storing of this program.

9.1.5.4 Double-Start Thread-Turning

Either G92 or G76 codes can be used to turn a double start thread or, for that matter, any number of starts. The method is to program the cycle as in our previous examples, cancel the G92 or G76 code by programming a G00 move with a Z-motion. The Z-motion will be one-half of the "*F*" or "*E*" (lead) in the positive direction for a double start, one-third of the "*F*" or "*E*" for a triple start, and $1/n$ for "*n* starts of thread.

Then repeat the thread turning cycle with the same dimensions in the following tape block.

If the thread has three starts, the thread-turning cycle has to be repeated three times. For "*n*" starts, the cycle will be repeated "*n*" times in the same program. Prior to repeating the cycle there must always be a rapid (G00) slide motion to cancel the previous cycle. This same tape block will also establish the start point for the next threading cycle.

9.1.5.5 Summary

Understanding of the canned cycles in this chapter will make it possible to write efficient part programs for the machining of complex parts. The programming steps and cycle patterns of the following cycles were discussed can be summarized as follows:

Canned Cycles for Machining Centers.   These are preprogrammed subroutines permanently stored in the CNC system's memory. Every canned cycle has a fixed pattern and tape format. Each canned cycle is activated by a specific *G*-code.

G80   Canned Cycle Cancel. When programmed, it will cancel a previously used canned cycle in the same or previous part program.

G81   Drill Canned Cycle. When used in a part program, it will rapid the tool point to an "*R*" level, feed to a "*Z*" depth, and return the tool point, in rapid motion, to the "*R*" level if a G99 code is programmed in the same tape block. Any number of holes may be drilled by programming the *X-Y* coordinates of the holes in the blocks following the G81 cycle. Equally spaced holes can also be drilled if the tape block contains the number of holes to be drilled in the "*L*" address.

G84   Tapping Canned Cycle. The subroutine combines the necessary tool motions, linear Z-axis and rotary spindle, required for the tapping of predrilled holes. The programmer must calculate the correct feed and speed values for a specific thread in terms of the linear displacement corresponding to each spindle revolution. The tool rotates, rapids to "R" level, feeds to "Z" level, reverses the spindle rotation and feeds back to "R" level, reverses the spindle to its original direction of rotation. The "L" address can also be programmed for equally spaced hole tapping to the same depth.

G73   High-Speed Peck-Drilling Cycle and G83   Peck Drilling Cycle. These are used for deep hole drilling. The cycle can be programmed to interrupt the feed motion to allow the swarf to clear any number of times. The cycle pattern is identical to the G81 with the exception of the feed motion.

G76   Fine Boring Cycle. This cycle is used for finish boring holes, with single point tools, to jig-bore accuracies. In the cycle pattern, the tool point will rapid to "R" level, feed to "Z" depth; the spindle will stop, orienting the tool point; the tool will shift by an amount "Q" (in direction opposite to the tool point) and will rapid to "R" or initial level.

G89   Boring Cycle. This cycle is used for spotfacing or counter boring applications.

Canned Cycles for Turning Centers.   These are divided into two groups: *canned cycles* normally control the tool point motions in a defined pattern for one cycle; *multiple repetitive cycles* will control the tool point through a number of cycles; while the shape of the pattern is maintained, the size of the motions is changing.

G90   Turning-Boring Cycle. This cycle is used for stock removing in both external and internal applications.

G90   Taper Turning Cycle. This cycle is used for rough turning of internal or external tapers. The tape format must contain the radial taper information in the "I" address. The sign of "I," depending on the taper, may be positive or negative.

G94   Facing Cycle. Like the G90 in turning, the G94 controls the tool point in facing applications. The magnitude of the taper is programmed under the "K" address; its sign may also be positive or negative. If the tape block with the G94 code has no "K" address, the format is for straight internal or external turning.

G71   Stock Removing Cycle. This cycle is used for rough turning of diameters with complex contours in several passes. The tape format defines the finished part's contour with allowances for finish turning. The contour, often called "part boundary," may be defined through linear and circular interpolation. This cycle should be used in constant surface speed (G96) mode.

G72   Stock Removal in Facing. This cycle is identical to the G71 code, but it is used for facing. The G72 is used when several facing passes are required to remove a large stock of material.

G70   Finish Turning Cycle. This cycle is normally used after the G71 or G72 cycles, within the same program. The part contour is defined in the stock removing cycle; therefore, the G71 code sends the control back to repeat tape blocks without the allowances left on diameters and faces.

G74   Multiple Grooving and Peck-Drilling Cycle. Using the same tape format, the G74 code can be utilized to control peck-drilling as well as turning grooves in the face of a part.

G75   Multiple Groove Turning Cycle. The G75 code controls the *X*-turret slide motions in feed and the *Z*-turret slide motions in rapid mode.

G92   Single-Start Multipass Thread-Turning Cycle. This cycle is used for turning straight and tapered internal and external threads.

G76   Multipass Multistart Thread-Turning Cycle. This cycle is used to program straight and tapered external or internal threads in one single tape block. Multistart threads can be turned by moving the start point of the turning tool by one-half lead distance for double start, one-third lead distance for triple start, and $1/n$ lead distance for "*n*" start threads.

EXERCISES

9.1. What are canned cycles?

9.2. Are canned cycles identical for all machine controllers?

9.3. Are all canned cycles fixed?

9.4. Specify the general format of the fixed canned cycle for a machining center and explain the codes.

9.5. Specify and explain the three levels of the machining center tool in Z direction.

9.6. What is the canned cycle cancel code, and when is it used?

9.7. What is the drilling canned cycle code, and how does the cycle work? Can it be used in all instances for drilling?

9.8. Does the canned cycle code have to be reprogrammed for each hole to be drilled?

9.9. What is the peck-drilling canned cycle code, and how does the cycle work?

9.10. What is the high-speed peck drilling canned cycle code, and what is the difference between it and the peck-drilling cycle?

9.11. What is the code for boring canned cycle, and how does the cycle work? What is the difference between it and the drilling canned cycle?

9.12. What is the code for fine boring canned cycle, and what is the difference between it and the boring canned cycle?

9.13. What is the code for tapping canned cycle and how does it work?

9.14. Student projects: Write programs for parts shown in Figure 5.14, p. 91, using canned cycles for center drilling (G81), peck drilling and tapping.

9.15. What is the code for straight-turning canned cycle and explain its operation.

9.16. Does the preceding code self-cancel?

9.17. Can the straight-turning canned cycle code be used for a different cycle as well? If yes, which one, and how?

9.18. Explain the facing canned cycle process, by comparing it to straight turning.

9.19. Explain the taper face turning canned cycle, by comparing it to the taper turning cycle.

9.20. What is a multiple repetitive cycle, when is it used, and what is its main advantage compared to a straight turning canned cycle?

9.21. Program the part from Figure 5.14, p. 91, using canned cycles as necessary.

10

Other CNC Features

10.1 SUBROUTINE PROGRAMMING

Subroutines, sometimes called subprograms, are a very powerful timesaving device. They provide the capability of programming certain fixed sequences or frequently repeated patterns and storing them in memory, and provide the opportunity of creating specific canned cycles for a particular application or product. The subroutines can be stored in memory under a specific program number and may be called up by the main program as needed.

The number of subroutines and their size is only limited by the total amount of storage capacity of the control. Subroutines are, in fact, independent programs, with their own program numbers, containing all the usual features of a stand-alone program. They are loaded in memory and deleted from memory just like any ordinary program, and, once stored, are viewed by the control in the same manner as a part program.

They have applications in turning and machining centers, as well as other CNC systems.

The subroutine (potentially one of several used in a program), may be called any time as required, and repeated any number of times. Following completion of the required instructions, the command may be returned to the "main" program in the immediately following line of program, or at any other program line specified by the programmer.

For example, the instruction

N0100 M98 P210

will transfer the command to program number 210, already stored in memory, and containing a given number of tape blocks.

The last block in the subroutine may read M99, following which the command is returned to the line of program immediately following N0100, or it may read, e.g., M99 P0180, at which time the command will be transferred to sequence No. 0180.

In both cases, for the control used in this example, the sequence tape word is designated by the letter P.

The content of the subroutine depends, besides the existing capabilities of the machine-control system, on the experience and the imagination of the programmer.

Some possibilities follow:

1. A complex hole pattern, requiring extensive machining at each hole location, e.g., spot drilling, predrilling, drilling, reaming, grooving (using a special-purpose quill-mounted grooving tool) and chamfering.

   A regular program would change the first tool, set up tool length compensation, perform the machining at all the locations, stop, change the tool and repeat for all the various operations required. Throughout the program, the pattern of hole locations will be observed six times, in this case. Placing the hole pattern in subroutine will mean writing a program containing the locations of all the holes, assigning it a program number, and using it as needed. The programmer will set up a "main" program, including only tool changes, spindle and coolant start and stop, as well as the corresponding Z-motions in the appropriate canned cycle. The "main" program will call up and switch to "subroutine" for all the X-Y motions. This will result in a far shorter and overall more efficient program, as well as a better utilization of control storage which affects control response time.

2. A complex contour requiring a number of roughing and finishing milling passes. Again, the entire contour may be located in a "subroutine" program, leading to a much shorter tape.

3. A number of special grooves in a shaft, for which a specific canned (or automatic repeat, or multiple repetitive) cycle may not be available. The groove is programmed in its entirety in subroutine, while the main program will only move the appropriate cutter to the starting point of the next groove.

For the programmer with BASIC or FORTRAN programming knowledge, these subprograms can also be nested, i.e., used repeatedly in a predetermined pattern of instructions. It is indeed one of the areas in which the selection of a solution largely depends on the imagination of the problem solver.

The following "common sense" notes may be used as a checklist in conjunction with subroutine programming:

- If the main program is in absolute (G90) and the subroutine uses incremental (G91) for its own good reasons, the programmer should reinstate G90 prior to returning the command to the main program.
- If a canned cycle is in effect in the main program and it is not required in subroutine, the latter should start with a canned cycle cancel code. Conversely, should a canned cycle be set up and used in subroutine, it should be canceled prior to returning to main.
- If a cutter diameter compensation or tool nose radius compensation is set up in main, and its transfer to subroutine leads consistently to an error message, the compensation should be set up in subroutine and followed, as prescribed, only by "motion" tape blocks.
- If a subroutine contains a program stop with the corresponding spindle and coolant stops, the latter should be restarted prior to returning to the main program, unless a tool change is involved.
- If the subroutine starts with a return home G28 code, or one is present in the main program return line, the control should be either set to G91 for G28 X0 Y0, or the intermediate point should be selected carefully according to the programming manual instructions.

Essentially, the above notes indicate that errors that would never be made when writing a straight continuous program are quite likely to crop up as the command switches back and forth between different programs. Consequently a more-careful-than-usual dry run is indicated. For example,
Main Program, No. 0001

:0001 N0010	
N0020	
N0030	
N0040 P910 M98	Command M98 will switch the main program to subroutine, whose own program number is 910, executes its instructions, then return to N0050
N0050	
N0060	
N0070 P910 L2 M98	As above, except for the fact that the subroutine instructions will be executed twice, due to L2, then return to N0080
N0080	
N0090	
.	
.	

.
.
.

N0200 M30   End of main program

Subroutine Program, No. 910

   :910 N0010 . . . .   As the programs under discussion are separate stand-
                                               alone programs, there will be no confusion due to
                                               similar sequence numbers

   N0020 . . . .
   N0030 . . . .

   .
   .
   .

   N0100 . . . . M99   M99 is the command that returns the subroutine to the
                                               main program, in the sequence immediately following
                                               the one that originated it

If N0100 had read N0100 . . . M99 P0090, the subroutine would have returned to
sequence N0090 in the main program. Sequence numbers, usually identified by the
tape word "*N*," are designated as "*P*" when used in conjunction with subroutine
instructions.

EXERCISES

10.1.1. What is the difference between a subroutine and a standard canned cycle?

10.1.2. Is the use of a subroutine limited to the program for which it was created?

10.1.3. What is the difference between a subroutine and a regular program?

10.1.4. Is there a limitation in the number of times a subroutine can be called?

10.1.5. Why are subroutines used?

10.1.6. Give two examples of typical applications for subroutines.

10.1.7. Give two examples of potential errors occurring in conjunction with the use of subroutines.

10.1.8. Give an example of a program line containing a subroutine call. Explain each word.

10.1.9. Give an example of a subroutine line returning control to the main program, in the line immediately following the subroutine call line.

10.1.10. Can a subroutine return control to a specific line, other than the one following the subroutine call?

10.2 SAFETY (CRASH) ZONE PROGRAMMING

Safety (crash) zone programming, known as *stored stroke limit* on the General
Numerics group of controls, is available for both machining and turning centers.

Programming a safe zone sets up an "invisible" electronic crash barrier, which stops the machine and outputs an alarm when an attempt is made to have a tool penetrate it. This feature is an invaluable capability, since a "crash" that occurs only in the control software is to be preferred by far to an encounter at 400–600 ipm between a tool and a chuck, or a mill and a clamp.

The principle is the same for both milling and turning. In turning, the program will set up an imaginary "box," whose base is a square of side X and whose length is given by Z as follows:

$$G22 \; X\_Z\_I\_K\_$$

X, Z, I, and K are all measured from home position (machine zero). X and Z are the coordinates of the corner of the box the closest to machine zero. I and K are the coordinates of the diagonally opposite corner of the box, the farthest from machine zero.

Due to software limitations, the above tape words are to be set in metric only, using trailing zero tape format, even if the rest of the program is in decimal point programming mode.

In milling, the format is similar:

$$G22 \; X\_Y\_Z\_I\_J\_K\_$$

where X, Y, and Z are the coordinates of the corner of the box the closest to machine zero, and I, J, and K, respectively, the farthest coordinates.

Once G22 is set up, any attempt to penetrate the zone will result in a machine stoppage, from which the operator has to back out manually, in the direction exactly opposite to the "offending" motion.

An alternate code is available, G23, which, when combined with a motion statement, will allow the tool to penetrate the safety "box."

There is no crash zone cancellation code. The programmer must, in the appropriate tape block, reset the safety zone off the surface of the table, outside of the normal travel limits of the tool. For example (Note: "Zero" machine and control).

```
N0010 G80 G40 G49 G20 G00 G90
N0020 G92 X0 Y0 Z4.0
N0030 G22 X10.0 Y-4.0 Z-4.0 I14.0 J-7.0 K-15.0
```

Sequence N0030 sets up an electronic "box," with its point the closest to machine origin at 10, -4 and -4 inches (X, Y, and Z), and its point the farthest from machine origin at 14, -7, and -15 inches (as given by I, J, and K)

```
N0040 Z0.5
N0050 G01 Z-0.1 F5.0
N0060 G00 X9.5 Y-4.1          The spindle is now close to the boundary
                                    of the electronic box
```

/N0070 G01 X10.1   If executed, this command will cause an
                                                    electronic crash

/N0080 G23 G01 X10.1   This is the same instruction as N0070, but
                                                    the motion will take place without an
                                                    electronic crash due to G23. The latter
                                                    allows penetration of the box, but does
                                                    not cancel it.

N0090 G28 X0 Y0 Z0

N0100 G22 X-1.0 Y1.0 Z1.0 I-2.0 J2.0 K2.0   Sequence N0100 resets the electronic box
                                                    off the machine table, for all practical
                                                    purposes canceling it, insofar as this
                                                    program is concerned.

N0110 M30

11

User Macros

The latest series of CNC systems have introduced a totally new concept of programming. This capability provides unique programming of automatic cycles for families of parts (group technology) in the Custom Macro environment (programming of variable commands for specific customer needs). Designed and developed to work within a modern Computer Integrated Manufacturing (CIM) environment, these advanced programmable controllers allow the user to create a higher grade of specialized functions.

In previous controllers, repetitive sequences defined as subroutines (or subprograms) could be called by the main program whenever required. Command was transferred to the address indicated in the call instruction, the subroutine was executed once or several times as requested, following which control was returned to the main program.

This program customizing process was expanded by the use of *macros*, a term generally understood to represent a group of instructions. These instructions use symbolic variables in lieu of numerical values. The variables will define data which may be either unknown or changeable at the time the program is written. Accordingly, unlike subroutines, user macro programming utilizes variables, allows for computation between variables, constants or variables and constants, as well as control instructions, such as conditional branching.

Variables are classified as *local*, *common*, and *system* variables, depending

on their function and application. A macro variable is composed of the symbol #
and a letter, such as *i*, *j*, etc. An example of a variable is #*i*. The letter may be
assigned a numerical value or an entire mathematical formula used to calculate,
for instance, a dimension resulting from a number of other given dimensions. In
some cases an integer constant (i.e., a regular nondecimal number) may replace
the letter in the variable. An example of the latter is #3, or #10, etc.

11.1 DEFINITION OF A MACRO VARIABLE

#*i* = [formula]   Example of definition of a variable by a given
formula, allowing one, for instance, to calculate
the total length of a part from the sum of its
components, or any other calculation that may
be required by a programmer.

Equivalence of macro variables

#*i* = #*j*   A new variable, #*i*, was introduced, and was
defined as being equal to #*j*, assumed known
from before.

Addition of macro variables

#*j* = #*i* + #*k*   Algebraic addition of variables #*i* and #*k*
#*j* = #*i* − #*k*   Algebraic subtraction of variable #*k* from #*i*
#*j* = #*i* OR #*k*   Logical (boolean) addition of variables #*i* and
#*k*
#*j* = #*i* XOR #*k*   Logical Exclusive OR-ing of variables #*i* and #*k*

Multiplication of macro variables

#*j* = #*i* * #*k*   Algebraic product of variables #*i* and #*k*
#*j* = #*i* / #*k*   Algebraic quotient of variables #*i* and #*k*
#*j* = #*i* AND #*k*   Logical product of variables #*i* and #*k*
(It should be noted that the logical operations are performed for each bit of
32 words.)

Macro representing functions

#*j* = SIN [#*i*]   Sine of variable #*i* expressed in degrees
#*j* = COS [#*i*]   Cosine of variable #*i* expressed in degrees
#*j* = TAN [#*k*]   Tangent of variable #*k* expressed in degrees
#*j* = ATAN [#*k*]/[#*i*]   Angle in degrees, whose tangent is the ratio of
variables #*k* and #*i*
#*j* = SQRT [#*k*]   Square root of variable #*k*

$\#k = \text{ABS} [\#i]$	Absolute value of variable $\#i$
$\#k = \text{BIN} [\#j]$	Variable $\#k$ is the binary equivalent of variable $\#j$ (which was expressed in Binary Coded Decimal, or BCD)
$\#j = \text{BCD} [\#k]$	Variable $\#j$ is the BCD equivalent of the variable $\#k$ (which was given in binary)
$\#j = \text{ROUND} [\#k]$	Variable $\#j$ represents the rounded-off value of variable $\#k$.

The function ROUND will round off values in two distinct ways:

1. In an arithmetic or conditional logical expression, the decimal number will be rounded off to the nearest unit. For example, $\#1 = \text{ROUND} [2.3452]$, $\#1$ will become 2.0; $\#2 = \text{ROUND} [3.1325]$, $\#2$ will become 3.0; and IF ($\#1$ LE ROUND $[\#2]$) GO TO 10, since $2.0 < 3.0$, control is transferred to sequence 10.

2. In a standard word address program, the result of the function ROUND is to round off the value to the least input increment of that address. For instance, if the least input increment of Y is 0.001 (which is usual in metric), and a dimension is given at a higher level of precision (e.g., $\#1 = 2.3452$), then the instruction GO1 Y ROUND $[\#1]$ will result in GO1 Y2.345. It should be noted that had we written GO1 Y$\#1$, the result would have been the same as the system would have handled the above operation.

11.2 COMMAND TRANSFER: PROGRAM FLOW STATEMENTS

GOTO is a command that may be used conditionally or unconditionally. Additional instructions such as DO, IF, WHILE, and END will be discussed as part of subsequent programming examples.

Under normal circumstances, a program is executed line by line, in order of line numbers. This order may be changed when the user macro program encounters a program flow statement. In this case, execution will jump to a nonsequential line, "n," if certain "conditions" are met.

Conditional GO TO statement.   The format of this statement is

$$\text{IF } [\langle\text{Condition}\rangle] \text{ Go To n}$$

Example

```
N100 IF [#10 GT 0.] GO TO 130
N110 #10 = #10 + 360
N120 #8 = ABS [#8]
N130 #18 = #6 * COS [#10]
```

The condition set in N100 states that if the value of variable #10 is greater than 0, then the statements N110 and N120 will not be processed, and control will be transferred to statement N130. Else, control proceeds as usual to the following block N110.

Logical Operations.   The relationships below can be used in the Conditional Arguments.

#i EQ #j	equal (=)
#i NE #j	not equal (≠)
#i GT #j	greater than (>)
#i LT #j	less than (<)
#i GE #k	greater than or equal to (⩾)
#i LE #j	less than or equal to (⩽)
AND	logical AND
OR	logical OR
NOT	logical NOT

Unconditional GO TO Statement.   The format of this statement is

GO TO n

Example

```
N100 IF [#10 GT 0.] GO TO 130
N110 #10 = #10 + 360
N120 #8 = −ABS [#8]
N125 GO TO 140
N130 #18 = #6 * COS [#10]
N140 #19 = #18 + #1
```

The GO TO statement in N125 will cause an unconditional divergence to the line number specified in the statement (in this example N140) every time that statement is executed. Execution will skip from N125 to N140 as this statement does not include a conditional statement. An unconditional Go To statement may also specify a backward transfer.

Iteration commands.   The format of this statement is DO i. It may be used to execute a single statement or a sequence of statements. When a series of statements are used, the programmer must also identify the END sequence numbers in the DO statement.

Example

```
WHILE [#10 LE #11] DO 1
#12 = #12 + #1
#13 = #18 * SIN [#12]
#14 = #18 * COS [#12]
```

```
#15 = #24 + #13
#16 = #25 + #14
G90 X#15 Y#16
#10 = #10 + 1
END 1
```

The DO 1 will start the loop or iteration, and the END 1 will finish the same loop. The system is capable of processing multiple loops, so long as they are nested as shown by the following rule:

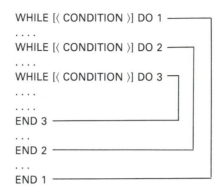

In the above example, DO 3 could be used for programming machining cycles (drill, tap, peck, etc). DO 2 could be used for line indexing, and DO 1 for row indexing.   The line–row indexing is handy for parts with multiple identical profiles.

The reader should be cautioned that while basic principles of programming are universal, specific differences apply from control to control, whether from different manufacturers, or even for the same make but different models, years, or option configurations.

Assignment of argument for variables.   When a variable command is programmed in a macro, the call statement in the main program assigns it a value, known as *argument*.

Address in main program	Variable in macro	Decimal point or digit position
A	# 1	3
B	# 2	3
C	# 3	3
D	# 7	0
E	# 8	0
F	# 9	0
H	#11	0
I	# 4	3
J	# 5	3
K	# 6	3

(Continued)

(*Continued from p. 181*)

Address in main program	Variable in macro	Decimal point or digit position
M	#13	0
Q	#17	3
R	#18	3
S	#19	0
T	#20	0
U	#21	3
V	#22	3
W	#23	3
X	#24	3
Y	#25	3
Z	#26	3

11.3 CUSTOM MACROS

The function of a custom macro is to establish a direct link between the computer and the machine tool by using a program. In conventional manual programming, the programmer has to perform all the calculations and place the results in the program. A computer-assisted programming language requires a description of the part and the tool path in a specific language. In custom macros, the programmer uses variables in formulas to specify the actual tool or slide movements, and these are calculated and implemented in the next line.

EXAMPLE

Write a part program for the milling of the repeating profile shown in Figure 11.1. The program will illustrate gradually advanced programming techniques. The tool is a 1/2 inch DIA center cutting end-mill, located at 0.1 inch over part zero.

11.3.1 Case (a): Subroutine Programming

It is assumed that the system has the M98 (go to subroutine) and M99 (return to main) programming functions.

O1000	*Program Number*
N10 G20 G40 G80 G90	Start-up, reset, set coordinate system.
N20 G28 G91 Z0.	Return tool axis to machine origin.
N30 G28 G91 X0. Y0. T01	Return tool to machine origin, index to tool No. 1.
N40 G92 X0. Y0. Z0.	New programmed zero at origin (fixture). Actual coordinate values to be edited in this line by operator.
N50 G90 S800 M3	Switch to absolute input, set spindle speed at 800 rpm and turn spindle on.
N60 G00 X-5. Y-4.37	Rapid to coordinate system X1-Y1 (see part sketch).

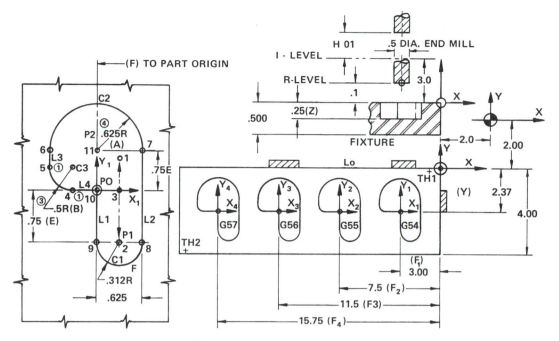

Figure 11.1   Repeating profile—sample part.

N70 G43 Z3. H01	Rapid to Z = 3.0″, set tool length compensation.
N80 G01 Z.1 F55. M08	Cut to R-level, coolant on.
N90 M98 P2000	Transfer control to subroutine (subprogram) 2000 to cut pocket 1.
N100 G00 X-4.5	Rapid to coordinate system X2-Y2 (2nd profile) at I-level on Z axis.
N110 M98 P2000	Transfer to subroutine 2000 to cut pocket 2.
N120 G00 X-4.	Rapid to X3-Y3 at I-level.
N130 M98 P2000	Transfer to subroutine 2000 to cut pocket 3.
N140 G00 X-4.25	Rapid to X4-Y4 at I-level.
N150 M98 P2000	Transfer to subroutine 2000 to cut pocket 4.
N160 G28 G91 Z0. M5	Return tool home on Z axis, stop spindle.
N170 G28 G91 X0. Y0.	Return tool home on X and Y axes.
N180 M30	End of program, memory rewind.
%	

O2000	*Subroutine Number*
N500 G92 X0. Y0.	Set part zero at location of tool (i.e., profile coordinate, X1-Y1 for the first profile, X2-Y2 for the second, etc.).
N510 G90 G00 X.3125 Y.3125	Rapid to 01 on R-level.
N520 G01 Y-.750 Z-.25 F5.	Cut to P1 and bottom of pocket.
N530 G45 Y0. D02	Cut to P3 with tool offset increase.
N540 G01 X-.125	Mill pocket to P4.
N550 G02 G46 X-.625 Y.5 R.5	Mill to P5.
N560 G01 Y.75	Mill to P6.

(Continued)

N570 G02 G48 X.625 R.625	Mill to P7.
N580 G01 Y-.75	Mill to P8.
N590 G02 G48 X0. R.3125	Mill to P9.
N600 G01 Y.25	Mill to P10 to P11.
N610 G00 Z.1	Return to R-level.
N620 G90	Absolute input.
N630 M99	Return to main program.

(NOTE: This solution presents a substantial reduction of programming effort. Subroutine No. 2000 was called 4 times, but it only appears in the program once. The direct saving represents 3 times 16 lines, hence an important time saving, while the indirect benefit is avoiding potential errors in 48 additional lines of program.

11.3.2 Case (b): Part Coordinate Setting

This case assumes that the CNC has a programmable multiple coordinate system feature, which would allow programming the locations of the profiles (X1-Y1, X2-Y2, etc.). The format is:

$$N... G10 \ Pn \ Xn \ Yn \ Zn,$$

where G10 represents offset value setting by program, Pn symbolizes the sequence of addresses P1, P2, etc. where the offset values will be assigned for X1, Y1, Z1, respectively X2, Y2, Z2, etc. These offset values are programmed using G54 . . . for P1, G55 . . . for P2, etc. Xn, Yn, Zn represent respectively the addresses X1, Y1, Z1, etc.

(NOTE: It is intended that this program reuse subroutine 2000 from case (a).)

O3000	*Program Number*
N05 G20 G40 G80 G91	Start-up, reset and set coordinate system.
N10 G28 G91 Z0.	Return tool axis to machine origin.
N15 G28 G91 X0. Y0. T01	Return tool to machine origin, index to tool No. 1.
N20 G92 X0. Y0. Z0.	New programmed zero at origin (fixture). Actual coordinate values to be edited in this line by operator.
N25 G90 S800 M03	Switch to absolute input, turn spindle on at 800 RPM.
N30 G00 G43 Z3. H01	Rapid to Z = 3.0″, set tool length compensation.
N35 G10 L2 P1 X-2. Y-2.37 Z0.	Assign these coordinate values for first pocket to the first register which will be activated by a G54 code.
N40 G10 L2 P2 X-7.5 Y2.37 Z0.	Coordinates for 2nd pocket to 2nd register, to be activated by G55.
N45 G10 L2 P3 X-11.5 Y-2.37 Z0.	Coordinates for 3rd pocket to 3rd register, to be activated by G56.
N50 G10 L2 P4 X-15.75 Y-2.37 Z0.	Coordinates for 4th pocket to 4th register, to be activated by G57.
N55 G54 G90 G00 X.3125 Y.3125 Z0.1	The G54 code will activate the preprogrammed values in

sequence N35 (location X1, Y1 and Z1), modified by the X, Y and Z values in the G54 block. This line will place the cutter at point 01, at R-level. However, this program design would require modifying the existing subroutine No. 2000. To avoid this, **THE SEQUENCE N55 WILL BE REPLACED BY ANOTHER ONE, WITH THE SAME NUMBER.**

N55 G54 G90 X0. Y0. Z0.	Preprogrammed values in N35 are not modified by this line.
N60 M98 P2000	Transfer to subroutine 2000 to cut pocket 1.
N65 G55 G90 X0. Y0. Z0.	Move to pocket No. 2.
N70 M98 P2000	Transfer to subroutine 2000 to cut pocket 2.
N75 G56 G90 X0. Y0. Z0.	Move to pocket No. 3.
N80 M98 P2000	Transfer to subroutine 2000 to cut pocket 3.
N85 G57 G90 X0. Y0. Z0.	Move to pocket 4.
N90 M98 P2000	Transfer to subroutine 2000 to cut pocket 4.
N95 G28 G91 Z0.	Tool axis return to machine zero about Z axis.
N100 G29 G91 X0. Y0.	Tool axis return to machine zero about X and Y axes.
N105 M30	Program end, memory rewind.
%	

11.3.3 Case (c): Custom Macro

The macro approach provides true computer-like programming with all the features necessary to perform the most complex calculations while the machine is cutting metal. Its primary advantage is that the programmer can design a customized program (specific canned cycles) for just about any complexity of machining, thus adapting the machine to the part requirements, rather than fitting parts to the controller's capability.

The program, in macro format, will be written after some additional discussion on macro program structure.

MAIN PROGRAM

```
N010 G20 G40 G80 . . .
N020 G28 G91 X0. Y0. Z0. T01
. . . .
. . . .
. . . .
N100 G65 Pn An Bn Cn . . . . Zn
```

If N100 is rewritten with numerical values, N100 G65 P9000 A3.25 B4.75 D6, then

G65—Go To Macro No. n (numerical example 9000)

A, B, C, . . . Z—addresses in main program

n—numerical values assigned to the addresses in the main program (numerical example 3.25, 4.75, 6)

(NOTE: Each of the above letter addresses has a corresponding macro number in the macro program, in accordance with the table in the section Assignment of Argument for Variables pp. 181–182.)

SUBPROGRAM MACRO

O9000	Macro Number
G28 G91 X0. Y0. Z0. . . .	
G01 X#1 F#7	Variable #1 = 3.25″; Variable #7 = 6 IPM

(From the variable table, main program address A corresponds to macro variable #1. In sequence N100, A was assigned the value 3.25 in main. The built-in table transposes this value to variable #1 in macro. Hence #1 = 3.25″. D was assigned the value 6 in main. The macro variable corresponding to D is #7, hence F#7 is the same as writing F6.).

Y#2	Variable #2 = 4.75″

(From the variable table, #2 corresponds to B, which was assigned the numerical value 4.75 in N100).

```
X-#1.
Y-#2.
G28 Z0.
 . . . . . .
```

CONCLUSIONS

1. This macro has milled a rectangular shape of 3.25″ (about X) by 4.75″ (about Y) at 6 IPM.
2. By changing the values of A and B in the main program (N100), any rectangular shape can be milled, within the limits of the machine, with no other changes required.
3. It should be noted that the macro subprogram 9000 does not contain any real numbers, as the variables #1, #2, etc. are symbols, equivalent to any letters used in algebra.

11.3.4 Overview of Variables

The variables acceptable to Fanuc type controllers are classified below.

Local variables. The range for these variables is from #1 to #33, and they include the values illustrated in the variable assignment table illustrated in this chapter. Local variables are the most common; their use, however, is limited to one macro within the same main program. For instance, if #33 is equal to 1.5 in macro A, the same #33 will *not* be equal to 1.5 in macro B. To facilitate visualizing the local variables, the following examples should be considered:

> If #33 = 1.5, then F#33 means a feed rate of 1.5 IPM in the block.
> If #18 = 6.2, then Z-#18 means −6.2″ tool axis motion.
> If #16 = 3, then G#16 means G03, circular interpolation CCW.

Various control options will offer different selections of local variables.

Common variables.   The usual range for these variables is from #100 to #149. Their use is also limited to one program; however, their range extends to all macros used by that program, not just one. If variable #33 = 28 in macro A, it will keep that value in macro B or C. This range of variables will, however, clear from memory when the power is turned off. A value calculated as follows

$$\#137 = \#6 * \sin [3.7 * 1.3]$$

in one macro can be used in all the other macros established by the same program only so long as power stays on.

Various control options will offer additional selections of common variables, some of which are not affected by power-off.

System variables.   The usual range for these variables is of values above #1000. They handle system-related functions such as links to other intelligent devices, tool offset amounts, work coordinate-system shift amounts, alarm message displays, RS232C data output, etc.

The range #1000 − #1015 is reserved for interface input signals. The range #1100 − #1115 is reserved for interface output signals. These system variables must always be on the right side of an equation in an operational expression (calculation). The range #2000 − #2299 is assigned to tool offsets, and can be used as follows:

> #10 = #2001 (the value of tool offset No. 1, located in variable #2001 is now assigned to variable #10, i.e., copied into #10).
> #8 = #2004 (as in the previous example, for offset No. 4, assigned to #8).

The range #2500 − #2506 is assigned to work offsets about the X-axis, and programmed using G54 to G59. The same way, the range #2600 − #2606 applies to Y and the range #2700 − #2706 applies to Z. For instance, #2605 = #3 means that the work coordinate offset for Y-axis, programmed using G58, is now set to be equal to the amount stored in #3.

The above concepts are among the most commonly used; however, other systems use different variables. Once the user understands the programming rules, it should be easy to move from one set of programming elements to the next one.

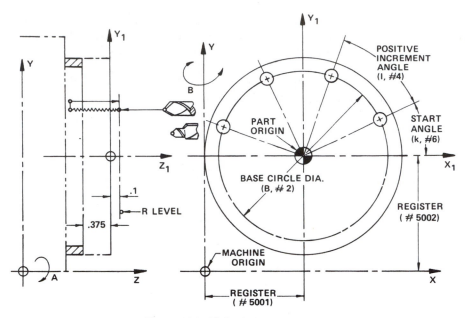

Figure 11.2   Hole circle pattern—sample part.

11.4 HOLE CIRCLE PATTERN APPLICATION—DRILLING MACRO

Macro techniques will be used to prepare the flexible drilling program shown below. The program was prepared using a five-axis machining center. In the case of a three-axis application, axes A and B would have no meaning and would have to be removed from the program. The same macro will service two different drilling canned cycles.

O4000	*Program Number*
N10 G17 G40 G20 G49 G80 G99	Start-up statement, specifying the X-Y plane (G17), canceling cutter diameter compensation (G40), tool length compensation (G49), and any prior canned cycles (G80). The tool is returned to R level (G99), and dimensions are in inches (G20).
N20 G91 G28 Z0.	Tool axis return to reference point—incremental input.
N30 G91 G28 X0. Y0. A0. B0.	Return tool axes to reference point.
N40 G92 X0. Y0. Z0. A0. B90	New programmed zero at origin. Actual coordinate values to be edited in this line by operator.
N50 T01 G90	Index to tool No. 1, center drill, and switch to absolute input.
N60 M06	Tool change implementation.
N70 S2000 M03 F5.5	Set spindle to 2000 RPM, turn spindle on, set feed.
N75 G00 B0.	Position part surface normal to drill centerline about B axis.
N80 G00 X3.0 Y3.0	Establish part zero in relation to machine zero. [NOTE: These values will be placed in registers #5001 (X) and #5002 (Y)

for variables #1 and #3 (see prog. O3100, seq. 10 and 20, below)]

N90 G43 Z3.5 H01 — Establish tool length compensation for the center drill.

N100 G81 Z-.25 R.1 L0 — Spot drilling canned cycle. Set drill depth. No motion will occur due to L = 0 condition (L—number of repeats).

N110 G65 P3100 B4.5 H4.I45.K25. [G65—"Go To" instruction, P3100—Subroutine number, programmed below as O3100. B, H, I, and K are main program addresses corresponding respectively to variables #2, #11, #4, and #6. The digits following B, H, I, and K are assigned as values to the respective variables.] — Transfer of control to macro (subroutine) O3100 and assignment of values to specified variables.

B (#2)—hole pattern base circle diameter (4.5″).
H (#11)—Number of holes = 4.
I (#4)—Incremental angle between holes = 45°.
K (#6)—Start angle = 25°.

The macro will generate the values for the center-drilling canned cycle.

N120 G28 G91 G80 Z0. — Return Z-axis to home.
N130 G28 G91 Y0. — Move Y-axis up to tool change position.
N140 G49 — Cancel tool length compensation.
N150 T02 G90 — Index tool magazine to position No. 2.
N160 M06 — Automatic tool change to ½″ diameter drill.
N170 S2200 M03 — Set spindle speed to 2200 RPM, turn spindle on.
N180 G00 X3. Y3. — See notes for N80.
N190 G43 Z3.5 H02 — See notes for N90.
N195 G00Z.1
N200 G73 Z-.55 R.1 Q.2 L0 — Peck-drill canned cycle.
N210 G65 P3100 B4.5 H4.I45.K25. — See notes for N110 (for peck drill holes center drilled).
N220 G28 G91 Z0. — Tool axis return about Z-axis.
N230 G28 G91 Y0. — Tool axis return about Y-axis.
N240 G28 G91 X0. A0. B0. — Tool axis return about X-, A-, and B-axes.
N250 M30
%

O3100 — *Hole Circle Pattern Drill Macro*

N10 #1 = #5001 — The X-dimension programmed at N80 will be assigned to variable #1.

N20 #3 = #5002 — As above, for Y-dimension.

N30 #5 = #2/2 — Assign value of "B/2" (4.5″/2 = 2.25″) from main program to variable #5.

N50 #7 = #11 — Assign number of holes, i.e., value of H = 4 to variable #7.

N60 #10 = #6 — Assign the value of start angle address, i.e., value of K = 25° to variable #10.

N70 #8 = #4 — Assign value of incremental angle, i.e., value of I = 45° to variable #8.

N80 #13 = 0 — Variable #13 will be used as a hole counter. Its value was set to zero.

(Continued)

N90 #22 = #4003	G90, set in N50, is assigned to variable #22. This may also be used for G91.
N100 IF [#10 GE 0] GO TO 130	If the start angle is greater than or equal to zero (as it is in this example), then the program will calculate the coordinates of the first hole, drill it, and repeat the steps until the condition set is satisfied. If the start angle #10 is negative, then the instructions in sequences No. 110 and 120 will be executed before sequence No. 130.
N110 #10 = #10 + 360	Increment negative start angle by 360°.
N120 #8 = ABS [#8]	Change sign of incremented angle.
N130 #20 = #5 * COS [#10]	The X-coordinate is the product of the radius and the cosine of its angle with the X-axis. This value is assigned to variable #20.
N140 #21 = #20 + #1	The previously calculated variable, #20, is added to the X dimension programmed at N80 (#5001).
N150 #18 = #5 * SIN [#10]	The Y-coordinate is the product of the radius and the sine of its angle with the X-axis. This value is assigned to variable #18.
N160 #19 = #18 + #3	The previously calculated variable, #18, is added to the Y dimension programmed at N80 (#5002).
N170 G00 X#21 Y#19	Move to X- and Y-coordinates of 1st hole, then execute the canned cycle programmed in the tape block prior to the macro statement.
N180 #13 = #13 + 1	Hole counter is incremented.
N190 IF [#13 EQ ABS[#7]] GO TO 220	Testing for completion of drilling; Has counter reached final value? If so, go to sequence No. 220.
N200 #10 = #10 + #8	Incremental angle is added to start angle.
N210 GO TO 130	Repeat calculation procedure for next hole.
N220 G80	Cancel canned cycles.
N230 G49 G#22	Cancel tool length compensation, restore G90/G91.
N240 M99	Return to main program.

(NOTE: This program can be changed to drill a different number of holes, starting from a different position, on a base circle of different diameter, by modifying the variables in the CALL ONLY statement.)

11.5 OUTSIDE CONTOUR MILLING APPLICATION—HEXAGON MACRO

The external hexagonal contour illustrated in Fig. 11.3 has been selected in order to exemplify the power and versatility of macro programming. This technique enables the programmer to produce an outside contour with any desired number of flat surfaces. The following assumptions are made to keep the problem within manageable limits:

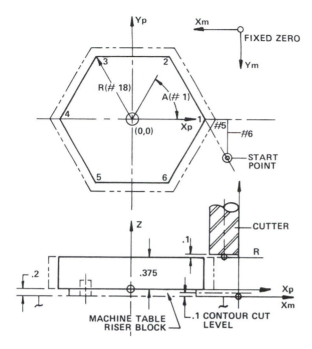

Figure 11.3 Outside contour(hexagon)—sample part.

- Use ⅝″ diameter mill.
- Part has been rough cut on bandsaw, maximum material not exceeding .20″.
- Part is held securely with a ½″-13 bolt, and rotation is prevented by a dowel.
- Part will be finish-milled with one pass around the contour.
- The CNC has programmable system variables.

O2500	*Program Number*
N10 G17 G20 G40 G80	Startup statement. Plane X-Y (G17), inch-dimensioning (G20), canceling prior cutter diameter compensation (G40), and canned cycles (G80).
N20 G28 G91 Z0.	Incremental input, tool axis pullout to home point.
N30 G28 G91 X0. Y0. T01	Return tool to reference point, incremental input, index to tool No. 1.
N40 G92 X0. Y0. Z0.	New programmed zero at origin. Actual coordinate values to be edited in by operator.
N50 M06	Tool change implementation.
N60 G00 G90 X0. Y0.	Absolute programming, rapid move to part zero.
G43 Z3. H01	Move tool to .1 above part.
N70 S1000 M3 M8	Set spindle speed, turn spindle on @ 1000 RPM, CW, turn coolant on.
N80 G65 P2501 R4. H6, X-12. Y-11. Z.375 Q0.3125 F5	Transfer of control to macro (subroutine) 2501, and assignment of values to specified variables:
G65—"Go To" instruction,	R [#18]—Outside radius of Hexagon.

P2501—Subroutine number programmed below as 02501. R, H, X, Y, Z, Q and F are main program addresses corresponding respectively to variables #18, #11, #24, #25, #26, #17, and #9. The numbers following R, H, X, Y, Z, and Q are assigned as values to the respective variables. (See table on pp. 181–182 for address and variable cross references).

H [#11]—Number of flats (sides).
X [#24]—X-distance from fixture to part origin.
Y [#25]—Y-distance as above.
Z [#26]—Z-distance as above.
Q [#17]—Tool radius = d/2.
F [#9]—Feed rate for milling.

N90 G28 G91 Z0.   Tool axis pullout.
N100 G28 G91 X0. Y0.   Return tool home.
N110 M30   End of program.

O2501   *Outside Contour Milling Macro*

#1 = 360 / #11   Variable #1 is designated to represent the center angle corresponding to a flat. It is calculated by dividing 360° by the number of sides (in this program, sequence N80 shows H, corresponding to variable #11, to be equal to 6—the number of sides). Therefore 360°/6 = 60°.

#2 = #26   Variable #2 is assigned the current value of Z (#26), i.e., .375, from seq. N80.

#3 = #4003   Variable #3 is assigned the current setting of either G90 or G91 code.

#4 = [180. − #1] / 2   The corner angle in the flat's isosceles triangle is calculated by subtracting the center angle (#1 = 60°) from 180°, and dividing by 2. (180° − 60°)/2 = 60°.

#5 = #18 + #17 * COS[#4]   Calculation of the X-address of the cutter's start location, by adding the X-projection of the tool radius to the part radius (radius of the circle that circumscribes the hexagon).

#6 = ABS [#17 * SIN[#4]]   Calculation of the absolute value of the Y-address of the cutter's start location.

#7 = #5   X starting coordinate from machine home.
#8 = − #6   Y starting coordinate from machine home.
G00 X#24 Y#25 M08   Rapid to part origin, coolant on. Dimension − 12. was assigned to X (#24) and − 11. to Y (#25) in sequence N80.

G92 X0.0 Y0.0 Z0.0   Program absolute zero origin at current location of tool.
#2015 = #17   Read the tool radius (Q, #17) into offset register No. 15.
G00 G91 G42 X#7 Y#8 D15   Rapid, incrementally, cutter diameter compensation on, to the X and Y starting coordinates, using tool offset register No. 15.

G01 Z0.0 F#9   Move to contouring level at 5 ipm per seq. N80 in program No. O2500.

#10 = 0	Set variable #10 to 0. This variable will be used as a counter in the loop below.
#12 = 0	Set variable #12 to 0.
N05 WHILE [#10 LE #11] DO1	Start loop No. 1 (DO1). Repeat the operations between this sequence and the end of loop 1 (see END1, below), so long as the value of counter #10, initialized above to 0, stays less than or equal to the number of flats, given above in #11.
N10 #12 = #12 + #1	In this sequence, variable #12 is reset to 60, as the current value of #12 is 0, and that of #1 is 60. The expression is equivalent to the statement #12 = 0 + 60.
N20 #13 = #18 * COS [#12]	The content of variable #13 is calculated as follows: Variable #18, R, was defined as 4.0 in seq. N80. #12 was just calculated to be 60, hence #13 = 4 * cos 60° = 2.00″.
N30 #14 = #18 * SIN [#12]	As immediately above, the calculation is 4 * sin 60° = 3.4641.
N40 G90 G01 X#13 Y#14 F#9	Mill flat 1-2 by moving cutter from the current (start) point to point No. 2.
N50 #10 = #10 + 1	Counter #10 becomes 0 + 1 = 1. Next time #10 will be used in the loop, its value will be one, until sequence 50 will be reached, when #10 will become 2, and so on.
N60 END1	End of loop 1. Control will be returned to sequence N05, and it will recycle until #10 will exceed #11, at which time the program will continue below. For details of each step of the subsequent loops, the calculations have been worked out at the bottom of the program.
G00 X#7 Y#6	Clear part.
G00 G49 Z#2	Rapid return to R level, cancel tool length compensation.
G00 G40 X0.0 Y0.0 M09	Rapid return to start point, cancel cutter diameter compensation, and turn coolant off.
G#3 M05	Restore G90/G91 setting and stop spindle.
M99	Return to main program.
%	

(NOTE: *The first loop* was outlined in the above programs. The subsequent loops will take place automatically as outlined in sequence N05. The following calculations will describe what exactly takes place in each of those loops.)

Second loop.   Coordinates for point P3, to machine flat from point 2 to point 3.

> N10 #12 = #12 + #1;   Last value for #12 was 60.   New value for #12 becomes 60 + 60 = 120.
> N20 #13 = #18 * cos[#12];   4 * cos 120° = −2.0.
> N30 #14 = #18 * sin[#12];   4 * sin 120° = 3.4641.
> N40 G90 G01 X#13 Y#14 F#9;   Cutter travel to point 3 will cause flat 2−3 to be milled.

N50 #10 = #10 + 1;   Increase counter #10 to 2, from its previous value which was 1. As 2 < 6, the looping process continues.

Third loop.   Coordinates for point P4, to machine flat from point 3 to point 4.

N10: angle will now index from 120° to 180°.
N20: 4 * cos 180° = −4.0.
N30: 4 * sin 180° = 0.0.
N40: Cutter travels to point 4, milling flat 3−4.
N50: Adding 1 to contents of counter will increase its value to 3. As 3 < 6, the process carries on.

Fourth loop.   Coordinates for point P5, to machine flat from point 4 to point 5.

N10: Angle becomes 180° + 60° = 240°.
N20: 4 * cos 240° = −2.0.
N30: 4 * sin 240° = −3.4641.
N40: Cutter travels to point 5, milling flat 4−5.
N50: Contents of counter become 4, and as 4 < 6, the procedure continues.

Fifth loop.   Coordinates for point P6, to machine flat from point 5 to point 6.

N10: Angle now increases to 300°.
N20: 4 * cos 300° = 2.0.
N30: 4 * sin 300° = −3.4641.
N40: Cutter travels to point 6, to machine flat 5−6.
N50: Contents of counter increase to 5 which is still within the limit, therefore, the looping process continues.

Sixth loop.   Coordinates for point P1, to machine flat from point 6 to point 1.

N10: Angle will index to 360°.
N20: 4 * cos 360° = 4.0.
N30: 4 * sin 360° = 0.0.
N40: Cutter moves to point 1, milling flat 6−1.
N50: Counter is now 6. The program has accomplished the condition set out in seq. N05. Control of the program is now allowed to continue on to sequence 60.

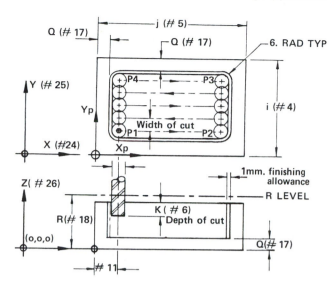

Figure 11.4 Inside contour (pocket)—sample part.

11.6 INSIDE CONTOUR MILLING APPLICATION—POCKET MACRO

The pocket milling application illustrated below will be programmed using macro programming techniques. The part dimensions will be expressed by utilizing variables. As these are assigned in a macro call line, the program will be easily adaptable for other pockets by simply changing the values assigned to these variables. The following assumptions have been made in order to keep the problem within manageable limits:

- The dimensions are given in metric.
- Cutter used has 12 mm diameter, center cutting.
- Finish cut allowance on the sides of pocket: 1 mm.
- Same wall thickness on all sides: 8 mm.

O2000	*Program Number*
N10 G21 G40 G80	Start-up statement. This includes canceling any prior cutter diameter compensation (G40) or canned cycles (G80). To switch this program to inches, replace G21 (metric) by G20 (inches).
N20 G28 G91 X0.0 Y0.0 Z0.0 T01	Return to reference point, incremental input, index to tool No. 1.
N30 G92 X0.0 Y0.0 Z0.0	New programmed zero at origin. Actual coordinate values to be edited in this line by operator.
N40 M06	Tool change implementation.
N50 G90 G00 X0.0 Y0.0	Absolute rapid move to part zero.

(Continued)

N55 G43 Z75 H01

Move 75 mm over part zero.

N60 S700 M03

Set spindle speed to 700 RPM, spindle on.

N70 G65 P9027 X25. I100. J200. Y0.0 Z0.0 K5. F200. D2. Q8. T60. R30.

Transfer of control to macro (subroutine) 9027, and assignment of values to specified variables: X [#24], Y [#25], and Z [#26] are absolute coordinates of the part coordinate system. I [#4]—width of part (outside dimension); J [#5]—length of part (outside dimension); K [#6]—depth of cut; F [#9]—feedrate for milling; D [#7]—tool offset register number (12 mm dia. becomes 6 mm radius); Q [#17]—wall and bottom thickness of part; T [#20]—rate of width of cut, expressed as a percentage of tool diameter, calculated as shown below:

G65—"Go To" instruction, P9027—Subroutine number programmed below as 09027. The numbers following the letter addresses are assigned as values to the respective variables (in metric). For instance, the outside length of the part, 200 mm, was assigned to the dimension J, corresponding in the program to variable [#5].

$$\text{Tool Dia.} \times \% \text{ of tool dia.} = \frac{12\text{mm} \times 60}{100} = 7.2 \text{ cuts}$$

R [#18]—rapid traverse level.

N80 G28 G91 Z0.0

Tool axis return.

N90 G28 G91 X0.0 Y0.0

Return tool home.

N100 M30

End of program.

O9027

Pocket Milling Macro Number

N10 #10 = #[2000 + #7]

Register #2002 contains the numerical value of the tool offset (i.e., 6 mm). 2002 results from 2000 + 2, (where 2 is the value assigned to D [#7] in N70). This value, 6 mm, is assigned to variable #10.

N20 #11 = #17 + 1 + #10

8 + 1 + 6 = 15 mm, where 8 mm (in this case the wall thickness) was assigned to variable #17 (Q) in N70, 1 mm is the finish cut allowance, and 6 mm is the cutter radius. This value, assigned to variable #11, sets up the X distance between the part origin and the start of the machining sequence.

N30 #12 = #4 − [2 ∗ #11]

100 − 2 ∗ 15 = 100 − 30 = 70 mm, where 100 was assigned to variable #4 (I) in N70, and 15 was calculated in the previous sequence. The value, 70 mm, assigned to variable #12, represents the internal width of the pocket.

N40 #13 = [#10 ∗ 2] ∗ [#20 / 100]

(6 ∗ 2) ∗ (60/100) = 12 ∗ 0.6 = 7.2 mm cut width, where 60 was assigned to variable #20 (T) in N70. This value, 7.2 mm, assigned to variable #13, represents the actual width of cut corresponding to each tool pass.

N50 #14 = FUP [#12 / #13]

70 / 7.2 = 9.7, rounded up to 10 cuts, as directed by the instruction "Fractional value to be rounded off **Up**wards." Therefore variable #14 will contain the actual number of passes. It will be used as the limit for the counter that will check for completion of the process. [NOTE: Because the use of sequence numbers is optional, they will be used below where their application is necessary (i.e., a reference will be made back to them).]

#27 = #24 + #11

25 + 15 = 40 mm, where 25 was assigned to the variable #24

	(X) in N70, and #11 was calculated above. The value, now stored in variable #27, represents the distance from the machine origin to starting point P1, along the X-axis.
#28 = #25 + #11	0 + 15 = 15 mm, where 0 was assigned to the variable #25 (Y) in N70. The value represents the distance from the machine origin to starting point P1, along the Y-axis, and is now assigned to variable #28.
#30 = #24 + #5 − #11	25 + 200 − 15 = 210 mm, where 200 was assigned to the variable #5 (J) in N70. The value represents the distance from the machine origin to end point P2, along the X axis, and is now stored in variable #30.
#15 = #4003	Read stored value of system variable #4003. This variable stores the status G90 or G91, in this case G90.
G00 X#27 Y#28 M08	Rapid traverse to starting point P1, above part.
G00 Z#18	Rapid to P1, R-level
#32 = #18	The value 30 mm, ascribed to variable #18 (R) in N70, is assigned to variable #32. This variable will be used in programming the succeeding depths of cut.
DO1	Start of the pocketing routine. First stage will be pocketing at the depth 5 mm, assigned to variable #6 (K).
#32 = #32 − #6	30 − 5 = 25 mm. This value represents the level of the first pass, measured from the (outside) bottom of the part.
IF [#32 GT #17] GO TO 60	So long as #32 (the current level of the pass) is greater than the part bottom thickness, 8 mm, the pocketing routine will continue as outlined at sequence N60, below. Control will be transferred to N60, and not to the next consecutive line of program.
#32 = #17	Set for last pass, i.e., #17 = 8 mm, bottom thickness.
N60 G01 Z#32 F#9	Cut from R-level of 30 mm to 25 mm height at 200 mm/min feedrate (value assigned to variable #9 [F] in seq. N70 in the main program). Of course, 25 mm correspond to the first pass. The depth values, contained in variable #32, will keep changing until the total depth has been machined.
G01 X#30	Mill (first) pass to point P2.
#33 = 1	Initialize a counter, using variable #33 for this purpose, to control the number of passes.
WHILE [#33 LT #14] DO2	Variable #14 was calculated above as the number of passes equal to 10 for this part. This loop will control the width of the pocket, by keeping the cutter moving along X between P1 and P2, and along Y by subsequent widths of cut, until the counter reaches the preestablished number of passes.
IF [#33 EQ #14] GO TO 70	If the condition of variable #33 (monitoring the number of passes) is equal to that of variable #14 (the total number of passes, in the case of this part 10), continue at sequence N70 in the macro program. This means that the pocket was

(Continued)

G01 Y[#28 + #33 ∗ #13]F#9

GO TO 80

N70 G01 Y[#25 + #4 − #11]

N80 IF [#33 AND 1 EQ 0] GO TO 90

completed on the current level. Else continue with the next line of program.

$15 + 1*7.2 = 22.2$ mm. This represents the second width of cut along Y-axis (initial Y + 1 times the width of cut).

Program continues at seq. N80

$0 + 100 − 15 = 85$ mm. Pocket width completed. Value calculated is maximum Y-dimension.

The cutter has reached the desired width of the pocket. This line of program tests for the cutter position at the left end or the right end of the pocket. In the first pass, and subsequent odd-numbered passes, the cutter traveled from left to right. In the second pass, and subsequent even-numbered passes, the cutter traveled from right to left. Sequence 80 states that as the cutter has reached the desired width, if variable #33 contains an even number, the pocket is completely finished, and control is transferred to sequence 90. If #33 contains an odd number, although the cutter is on the final width, it is in the right corner, and it has to return to the left, which will be implemented by the immediate following instruction (G01 X#27, below). The instruction works as follows: From basic logic, 0 AND 0 = 0; 0 AND 1 = 0; 1 AND 0 = 0; and lastly, 1 AND 1 = 1. For #33 containing even numbers, the condition stated in N80 is met. The numbers could be, in decimal and binary

2	0010
4	0100
6	0110
8	1000
10	1010, etc.

Any one of these numbers, "ANDed" with 1 (0001) will equal 0, as stated in N80, e.g.,

$$0010$$
$$\text{AND}$$
$$\underline{0001}$$
$$0000, \text{i.e.} \ldots \text{EQ } 0.$$

Conversely, if the contents of #33 are odd, as in

1	0001
3	0011
5	0101

etc., any one of the latter, "ANDed" with 1 (0001) will equal 1, e.g.

$$0001$$
$$\text{AND}$$
$$\underline{0001}$$
$$0001$$

i.e., . . . EQ 1.

G01 X#27	Mill from point P2 to point P1 along the X-axis (last pass).
GO TO 100	No more milling along the X-axis on this level.
N90 G01 X#30	Mill last pass to point P2 along the X-axis.
N100 #33 = #33 + 1	Increment the counter.
END2	End of the inner loop started at DO2. This loop controlled the tool path within the boundaries of the pocket.
G00 Z#18	Rapid tool to R-level.
IF [#32 LE #17] GO TO 110	If the depth level stored in #32 (currently 25 mm) is less than or equal to #17 (8 mm), then the pocket roughing has been completed and the next step is the finishing cycle, starting at N110. If the pass level has not reached 8 mm, continue with the next sequence.
G00 X#27 Y#28	Rapid to P1 on R-level for second depth rough pocketing.
G00 Z[#32 + 1.]	Rapid to #32 (25 mm) + 1 mm = 26 mm.
END1	From this point, the program will return control to the DO1 statement above, and will reset the depth to the second depth as in #32 = #32 − #6 (i.e., #32 = 25 − 5 = 20 mm). It will then loop through again and again, until the condition expressed as IF [#32 LE #17] is no longer satisfied. When #32 becomes greater than #17, control will be transferred to N110 below. (NOTE: Subsequent steps will set up the finish milling.)
N110 #11 = #11 − 1	15 − 1 = 14 mm. The value 15 mm for #11 was calculated in sequence N20 at the beginning of the macro program.
#27 = #27 − 1	40 − 1 = 39 mm, to the X start point in the pocket.
#28 = #28 − 1	15 − 1 = 14 mm, to the Y start point in the pocket.
#30 = #30 + 1	210 + 1 = 211 mm, to the X end point in the pocket.
#31 = #25 + #4 − #11	0 + 100 − 14 = 86 mm, to the Y end point in the pocket.
G00 X#27 Y#28	Rapid to start point in pocket.
G01 Z#17 F#9	Feed to 8 mm, bottom of pocket. This value was assigned to #17 (Q) in N70 in main program.
X #30	Finish mill inside pocket to P2 along the X-axis.
Y #31	Finish mill inside pocket to P3 along Y-axis.
X #27	To P4 along X.
Y #28	Back to P1 along Y.
N130 G00 Z#18	Rapid to R-level.
G00 X#24 Y#25	Rapid to start point.
G#15	Restore G90/G91.
N150 M99	Return from subroutine to main program.
%	

(NOTE: This program can be changed to any pocket size by modifying the variables in the CALL statement. Unequal wall thicknesses would require the introduction of additional variables.)

EXERCISES

11.6.1. Describe the meaning of the following statements:
1. IF [#5 EQ #31] GO TO 25
2. IF [#9 LE #30] GO TO 31
3. IF [#8 LT #32] GO TO 35
4. IF [#7 GE #33] GO TO 36
5. IF [#10 GT #20] GO TO 40

11.6.2. Given the following portion of a program,

```
#13 = (see below)
#12 = 2
DO1
IF [#12 GT 6] GO TO 10
#12 = #12 + #13
END 1
. . . .
. . . .
N10 . . . .
. . . .
```

determine the number of times the program will cycle between DO1 and END 1 if:
1. #13 = 2
2. #13 = 3
3. #13 = 4
4. #13 = 5
5. #13 = 0
6. #13 = 6

11.6.3. Given the following portion of a program,

```
#13 = (see below)
#21 = (see below)
#12 = 7
WHILE [#12 LT #13] DO 2
#12 = #12 + #21
. . . .
. . . .
. . . .
END 2
. . . .
```

determine the number of times the program will cycle between WHILE and END 2 if:
1. #13 = 11 and #21 = 1
2. #13 = 14 and #21 = 2

    3. #13 = 12 and #21 = 10
    4. #13 = 15 and #21 = 3

11.6.4. Design a program to produce a rectangular pocket (slot) at any angle with respect to the X-axis, a vertical pocket, a horizontal pocket and a circular pocket. The various results are to be achieved by changing the variables in the CALL statement.

11.7 FAMILY OF PARTS PROGRAMMING—LATHE MACRO

The "family of parts" programming concept is one of the more efficient means to reduce the required number of part programs and improve the productivity of the CNC operations. This approach should be considered when the parts to be machined are similar in appearance and within the dimensional capabilities of the CNC lathe to be used. In the following example, four different components, similar in shape, are considered.

    Task.   Write a User Macro part program for the machining of the family of parts shown in Figure 11.5.

    Assumptions.   The CNC lathe has user macro programming capability. The controller will be able to perform arithmetic operations, accept and use variables, and perform logical operations. The CNC will have the capability to function as a computer, and at the same time as a machine motion and function controller.

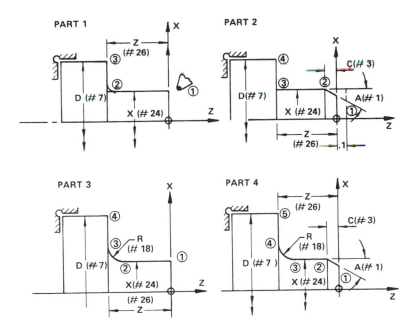

Figure 11.5 Family of parts—lathe macro.

Solution. A macro was designed to search for the part type, perform all the necessary calculations, and direct the machine controller to turn the appropriate part configuration.

The main program will have to perform the usual start-up procedure, establish the part-machine coordinate relationship and identify the variables for a specific part type of the family. The decision on how to write this section may vary by application or user. *If the shop personnel are trained and able to edit an existing program* (i.e., delete or replace variables), *then a single program is sufficient.* These changes apply to the call transfer statement in the main program, as follows:

- PART 1

      N80 G65 P7777 A0. C0. D2.25 R0. X1.75 Z2.5

- PART 2

      N80 G65 P7777 A30. C0.3 D2.25 R0. X1.85 Z2.48

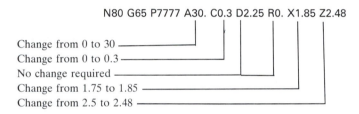

Change from 0 to 30
Change from 0 to 0.3
No change required
Change from 1.75 to 1.85
Change from 2.5 to 2.48

- PART 3

      N80 G65 P7777 A0. C0. D2.25 R0.07 X1.90 Z2.48

  Note changes required for A, C, R and X

- PART 4

      N80 G65 P7777 A35. C0.1 D2.25 R0.1 X1.96 Z2.40

If this method is chosen, one must make a first part inspection to assure that the proper values were inserted into the program.

O6001	*Family of four parts. Part numbers to be specified.*
N10 G20 G40 G99	Inch programming, cancel tool nose radius compensation, feed per revolution.
N20 G50 X5.0 Z8.0 S1200	Programming of absolute zero point, X and Z distances from part to machine zero. When the speed word is programmed in the same block, this will set the maximum spindle speed for this program.
N30 G00 T0200 M42	Index to Tool No. 2, spindle speed for gear range No. 2— per M-code table.

N40 G96 S700 M03

Set constant surface speed to 700 FPM, turn spindle on CW.

N50 G00 G42 X2.5 Z0.3 T0202 M08

Tool nose radius compensation right, rapid to programmed point 1 with offset register No. 2, turn coolant on.

N60 G65 P7777 A0. C0. D2.25 R0.
    X1.75 Z2.5

G65—user macro transfer command to subprogram No. 7777, describing the variables required to turn PART 1.

A[#1] Angle $= 0°$
C[#3] Chamfer $= 0''$
D[#7] Blank Diameter $= 2.25''$
R[#18] Radius $= 0$
X[#24] Turned Diameter $= 1.75''$

The same block could have been programmed as:

N60 G65 P7777 D2.25 X1.75 Z2.5

since variables of numerical value zero can be omitted from the program.

N70 G00 G40 X5.0 Z8.0

Cancel tool nose radius compensation, rapid to clear part after the control has returned from the subprogram (NOTE: this is the line after the call, to which the program control returns at the end of the subprogram).

N80 T0200
N90 M30

Reset memory to start point

%

End of program.

O7777
N10 #100 = [(#7 − #24) / 2]

User Macro for Family of Four Parts

Calculate depth of cut as one-half the blank diameter less the turned diameter. [#7 (2.25″) − #24 (1.85″)]/2 = 0.2″.

N20 IF [#18 LE #100] GOTO 40

Test for radius condition. If the radius (#18) is less than or equal to depth of cut 0.2″ (#100), the part cannot be turned, and control is transferred to sequence N40.

N30 #3000 = 199

Turn alarm message 199 on: Radius too large!

N40 #101 = [#3 + 0.1] * TAN[#1]

For PART 1 and PART 3, #3 = 0 and #1 = 0, therefore #101 = (0 + 0.1)*Tan 0 = 0. For PARTS 2 and 4, variable #101 will have to be calculated based on the assigned values for #3 and #1.

N50 G00 X[#24 − 2*#101] Z0.1

Rapid to start point 1. Since #101 = 0, 2 * #101 = 0, as a result this block will become N50 G00 X1.75 Z0.1, as #24 = 1.75″.

N60 IF [#101 LE 0] GOTO 80

Since #101 = 0, we go directly to seq. N80, bypassing seq. N70, which handles chamfer turning.

N70 G01 X#24 Z-#3 F0.15

For PART 1, this block will not be used because the tool point is at point 1. Before the next turning block is programmed, the *presence* or *absence* of the radius must be established.

N80 IF [#18 LE 0] GOTO 120

If there is a radius, the subsequent lines will guide the tool point to points 2, 3, and 4 for PART 3 and 2, 3, 4 and 5 for PART 4 as follows:

(Continued)

N90 G01 Z-[#26-#18]   Turn #24 diameter to start of radius, point 2 on PART 3
                                                       and point 3 on PART 4.
N100 G02 X[#24 + 2*#18] Z-#26   Turn radius, X = #24 + 2*#18 = 1.75 + 2*0.1 = 1.95
       R#18
N110 GOTO 130   Bypass seq. N120, which deals with the R = 0 situation.
N120 G01 Z-#26 F0.01   If there is no radius, the tool point will cut to point 2 for
                                                       PART 1 and 3 for PART 2.
N130 X[#7 + 2*0.1]
N140 M99   Return from subprogram.
%   End of subprogram.

(NOTE: If the necessary know-how is not available on the shop floor, a separate
program can be written for each part in the family, as shown below. Program O6001
is already available from p. 202.)

O6002   *Program for PART 2 of the Family*
N10 G20 G40 G99   Inch programming, cancel tool nose radius
                                                       compensation, feed per revolution.
N20 G50 X5.0 Z8.0 S1200   Programming of absolute zero point, X and
                                                       Z distances from part to machine zero.
                                                       When the speed word is programmed in
                                                       the same block, this will set the maximum
                                                       spindle speed for this program.
N30 G00 T0200 M42   Index to Tool No. 2, spindle speed for gear
                                                       range No. 2—per M-code table.
N40 G96 S700 M03   Set constant surface speed to 700 FPM, turn
                                                       spindle on CW.
N50 G00 G42 X2.5 Z0.3 T0202 M08   Tool nose radius compensation right, rapid to
                                                       programmed point 1 with offset register
                                                       No. 2, turn coolant on.
N60 G65 P7777 A35. C0.15 D2.25 X1.75 Z2.5   G65—user macro transfer command to
                                                       subprogram No. 7777, describing the
                                                       variables required to turn PART 2.
N70 G00 G40 X5.0 Z8.0   Cancel tool nose radius compensation, rapid
                                                       to clear part after the control has returned
                                                       from the subprogram (NOTE: this is the
                                                       line after the call, to which the program
                                                       control returns at the end of the
                                                       subprogram).
N80 T0200
N90 M30   Reset memory to start point.
%   End of program.

O6003	*Program for PART 3 of the Family of Parts*
N10 G20 G40 G99	Inch programming, cancel tool nose radius compensation, feed per revolution.
N20 G50 X5.0 Z8.0 S1200	Programming of absolute zero point, X and Z distances from part to machine zero. When the speed word is programmed in the same block, this will set the maximum spindle speed for this program.
N30 G00 T0200 M42	Index to Tool No. 2, spindle speed for gear range No. 2—per M-code table.
N40 G96 S700 M03	Set constant surface speed to 700 FPM, turn spindle on CW.
N50 G00 G42 X2.5 Z0.3 T0202 M08	Tool nose radius compensation right, rapid to programmed point 1 with offset register No. 2, turn coolant on.
N60 G65 P7777 D2.25 R0.202 X1.75 Z2.5	G65—user macro transfer command to subprogram No. 7777, describing the variables required to turn PART 3.
N70 G00 G40 X5.0 Z8.0	Cancel tool nose radius compensation, rapid to clear part after the control has returned from the subprogram (NOTE: this is the line after the call, to which the program control returns at the end of the subprogram).
N80 T0200	
N90 M30	Reset memory to start point.
%	End of program.
O6004	*Program for PART 4 of the Family of Parts*
N10 G20 G40 G99	Inch programming, cancel tool nose radius compensation, feed per revolution.
N20 G50 X5.0 Z8.0 S1200	Programming of absolute zero point, X and Z distances from part to machine zero. When the speed word is programmed in the same block, this will set the maximum spindle speed for this program.
N30 G00 T0200 M42	Index to Tool No. 2, spindle speed for gear range No. 2—per M-code table.
N40 G96 S700 M03	Set constant surface speed to 700 FPM, turn spindle on CW.
N50 G00 G42 X2.25 Z0.3 T0202 M08	Tool nose radius compensation right, rapid to programmed point 1 with offset register No. 2, turn coolant on.
N60 G65 P7777 A60. C0.2 D2.25 R0.19 X1.75 Z2.50	G65—user macro transfer command to subprogram NO. 7777, describing the variables required to turn PART 4.

(Continued)

N70 G00 G40 X5.0 Z8.0   Cancel tool nose radius compensation, rapid
                                                   to clear part after the control has returned
                                                   from the subprogram (NOTE: this is the
                                                   line after the call, to which the program
                                                   control returns at the end of the
                                                   subprogram).
N80 T0200
N90 M30   Reset memory to start point.
%   End of program.

Each of these four programs use the same identical format for start-up. They all use *a common subprogram*, which achieves a substantial saving in programming. The principles used in this program, for the system described, are easily transferred to other CNC systems.

EXERCISES

11.7.1. Calculate the value of #101 for PART 3 if #3 = 0.2 and #1 = 37.5.

11.7.2. Calculate the value of #101 for PART 4 if the angle A = 60° and the chamfer C = 0.37″.

11.7.3. Calculate the actual X- and Z-coordinates of point 1 for the above questions if the values of #24 and #26 are the same as shown for PART 2.

11.7.4. Write down the order of sequence numbers controller will follow in the user macro program for PART 1, PART 2, PART 3, and PART 4.

11.7.5. Calculate the Z-coordinate for tool location 3 and the X-coordinates for tool location 4 for PART 4 if #24 = 3.8, #26 = 2.65, and #18 = 0.185.

11.8 SLOT MILLING APPLICATION—CINCINNATI MILACRON 850/950

The Cincinnati Milacron 850/950 controls use the same principles described in the previous applications. The differences are only in appearance, and can be summarized as follows:

- The subroutine (macro) starts with the acronym DFS (**DeFi**ne **S**ubroutine), followed by a program number. This number will be called from a main program.
- The subroutine ends with the acronym ENS (**EN**d **S**ubroutine).
- The call statement in the main program should contain at least a sequence number, the acronym CLS (**CaL** **S**ubroutine), and the letter L followed by the number of the defined subroutine. The latter can range from 0 to 499. All the other standard word addresses may appear in this line as required.

- Variable information is passed through specific or temporary variables, and up to 18 words may be included in the subroutine to define the subroutine variables. In lieu of "#" followed by a number, this control system uses "[P]" followed by a number, as shown in the table below:

Variable reference	Parameter retrieved
[P1]	G
[P2]	X
[P3]	Y
[P4]	Z
[P5]	B
[P6]	I
[P7]	J
[P8]	K
[P9]	F
[P10]	S
[P11]	T
[P12]	M
[P13]	R
[P14]	A
[P15]	C
[P16]	U
[P17]	V
[P18]	W

As in the previous examples, when a word address is not programmed in the call statement, the corresponding P variable is set to zero for the subroutine. The subroutine cannot change the numerical values of P.

Task.   Write a program to machine four slots in the part shown in Figure 11.6.

Assumptions.   0.5″ diameter end mill. For simplification, the main program will only show the portion including the subroutine call statements. No canned cycles will be shown. The subroutine will be defined in the first part of the program listing below. Note abbreviation MSG for message.

Solution

(MSG, SUBROUTINE FOR MILLING SLOTS)	Operator information.
(DFS,10)	Start definition of subroutine, program No. 10.
G0 X1 + [P2] Y1 + [P3] Z.850 S685 F12 M3	Rapid to position 1, 0.1″ above part.
G1 Z-.1	Feed through the part.
X1 + [P2]-.125 Y1 + [P3] + .4	Feed to position 2.
Y1 + [P3] + 1	Feed to position 3.

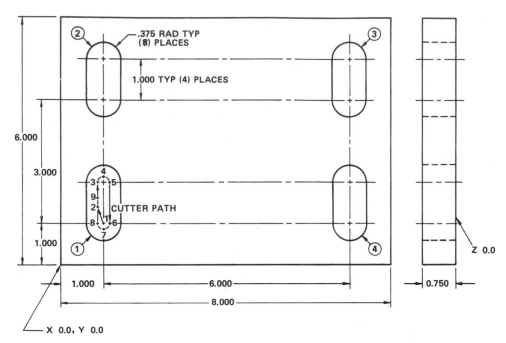

Figure 11.6 Slot maching—sample part.

G2 X1+[P2] Y1+[P3]+.125 I1+[P2] J1+[P3]+1	Feed to position 4.
X1+[P2]+.125 Y1+[P3] I1+[P2] J1+[P3]+1	Feed to position 5.
G1 Y1+[P3]	Feed to position 6.
G2 X1+[P2] Y1+[P3]-.125 I1+[P2] J1+[P3]	Feed to position 7.
X1+[P2]-.125 Y1+[P3] I1+[P2] J1+[P3]	Feed to position 8.
G1 Y1+[P3]+.6	Feed to position 9.
G0 Z.850	Retract to .100 above part.
(ENS)	End subroutine definition.
(MSG, PART PROGRAM)	
"	
"	
"	
(MSG, LOAD .5″ DIA. ENDMILL)	
G0 T1 M6	Load tool No. 1.
CLS, L10 X0 Y0	Call subroutine for lower left slot, No. 1.
CLS, L10 X0 Y3.	Upper left slot, No. 2.
CLS, L10 X6. Y3.	Upper right slot, No. 3.
CLS, L10 X6. Y0.	Lower right slot, No. 4.
"	
"	
"	

In the subroutine, the variables [P2], [P3] or [P4], if applicable, are given the values outlined in the call instructions. For the above program, these can be summarized as follows:

Call Number	Variable Equivalent	[P2] X	[P3] Y	[P4] Z
1st call		0	0	0
2nd call		0	3	0
3rd call		6	3	0
4th call		6	0	0

Accordingly, a program instruction X1 + [P2] -0.125 will represent $(1 + 0 - 0.125)''$ after the first or second call, but $(1 + 6 - 0.125)''$ after the third or fourth one.

As seen above, the subroutine calls are identical, except for the address of the slot, and the same set of instructions carries out the slot machining in all four cases. The subroutine, though, could be defined in far more general terms and used for a number of different applications. Up to three levels of nesting can be used with these controllers.

12

Computerized CNC
Part Programming

In the previous chapters we have explored the many intricacies of manual part programming. The term *manual* applies to all situations in which the program is written in machine codes, even though microcomputers may have been used to prepare and process the tape. The term *computer-assisted* represents the use of a specific language, to be discussed later in this chapter.

In spite of the efforts of the manufacturers and the various professional organizations, the CNC industry has so far been unable to come up with a compatible coding standard. This was primarily because no national or international body has the power to regulate or enforce the uniform use of codes for manual programming. As a result, the same "G" code may represent three different functions on three different systems. Programs written for one control type cannot be used for another, even if the machine is the same kind and size, because of incompatible codes. The authors have encountered incompatibility even when the machine and control were the same, because additional features had been built into one of the systems, somewhat more recent than the other. For this reason, programs often had to be rewritten—at added cost—if jobs had to be shifted from one machine to another. Alternatively, the company loses flexibility in the area of shop scheduling.

In addition to coding, manual part programming requires accurate and consistent calculations, occasionally far more extensive than the coding itself.

In Chapter 11, a recent concept of macro programming was introduced and discussed, involving additional codes and programming rules and providing extensive flexibility and efficiency. This user-written subroutine (macro) programming requires additional knowledge in trigonometry and analytical geometry. It also, unfortunately, carries the same burden as its predecessors: the programming formats for the mathematical elements are incompatible.

Manufacturers introduce new programming features yearly, independent of each other, and the result is that the same problem is programmed differently on each control. A programmer who programs manually several totally different machine-control systems is prone to make costly mistakes in terms of scrapped parts, broken tools, and machine damage. These situations delay production schedules and add to the cost of the manufactured product. Before getting involved in our discussion of computer-assisted systems, we should emphasize that manual programming is performed effectively and efficiently for two-axis contouring applications. Relatively simple parts, with linear and circular boundaries, can be programmed with ease, while more complex parts, such as cam contours, will require a working knowledge of analytical geometry.

Once the calculations are done, the programmer must combine the calculated data with the appropriate codes, acceptable to the respective CNC system, to produce a working part program.

The programmer's task is difficult, but possible. This task becomes harder with full three-axis contouring, and impossible for four- and five-axis applications.

The Computer-assisted programming does offer a number of significant benefits as an upgrade from manual programming. The computer can now run this program to produce a tape, file, or CNC program for the machining of a part. It is however a step back from the design, as the programmer must translate each element of the drawing, (point, line, arc, circle, etc.) into a mathematical definition or a computer statement. An already dimensioned and toleranced part drawing must be redefined into numbers, thus wasting a lot of the designer's work, instead of being defined using the same graphic methods that led to its conception.

The speed and accuracy of the computer produce tapes or programs with significantly less lead time, without errors. Most computer-assisted systems allow the programmer to plot part geometry and the tool path with significant accuracy, thus cutting down on expensive machine time requirements for tape tryouts. The computer routinely transforms the calculated slide motions to the correct tape format, thus allowing the programmer to concentrate on the program, not on the features of the system on which the part is to be machined.

The benefits of computer-assisted programming are most significant in terms of programming time, accuracy, and flexibility, which can transform a CNC operation into a money-making concern.

12.1 COMPUTERIZED SYSTEMS FOR PART PROGRAMMING

The development of computer-assisted programming languages started in parallel with the production of the manufacturing systems. The manufacturers of CNC systems, as well as the first users, realized that three- four- and five-axis systems can only be utilized to their fullest capabilities if a workable programming language is available to the users. This language had to be simple enough for the users to learn, and at the same time complex enough to address all the capabilities of the CNC system. The first such language, called Automatically Programmed Tools (APT) was developed by MIT and became the most powerful full five-axis programming language.

In parallel with the growth of the CNC industry, numerous less complex computer-assisted languages were developed. Depending on the complexity of parts, as well as many other criteria, the users have a wide range of languages to choose from. Most of these share a general system concept. All computer-assisted CNC programming languages use "English-like" words to describe the part, tool, type of operation, and the required auxiliary operations for the machining process. In some cases, the words may be abbreviated, sometimes to single letters. The CNC program, using the programmed data, performs the mathematical calculations and compiles a "cutter location" list or file. The processed data is then recycled through another CNC program called "postprocessor" or "link." The function of the latter is to change the processed data into coded tape formats acceptable to a specific machine-control system.

The CNC program is universal. It is the same for all the different CNC machines, which certainly eases the task of the programmer. The postprocessor, on the other hand, is a specific, special-purpose program, written for a well-defined machine-control system. The postprocessor knows all the programmable codes (G, M, T, S, etc.) and acceleration and deceleration characteristics, ASCII or EIA tape formats, in addition to speeds, feeds, and physical limitations of slide motions. It is a "link" between the CNC computer and the CNC machine. A schematic representation of an CNC computer-assisted process is illustrated in Figure 12.1.

The figure illustrates a typical CNC system. The program may be punched on tape and read into the MCU using a high-speed tape reader. This could be a short tape file keyed in through MDI or directly downloaded from the computer to the CNC. Once the program has been entered, it must be edited for errors. The process is performed by the programmer through an interactive terminal. The programmer should obtain a "list" or "CL print" as well as a tape file printout prior to having the machine tape punched out. If the system is equipped with a plotter, a part outline and cutter centerline path should be plotted.

The programmer must check all these data for accuracy and validity and make the final corrections where necessary, before the tape file is transferred to CNC.

There are so many computer-assisted CNC languages on the market that no

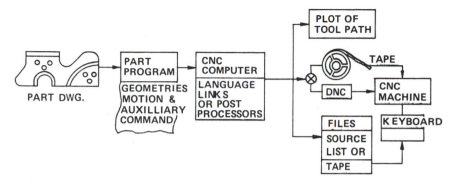

Figure 12.1   CNC computer system.

purpose would be served in listing even the most common ones. New languages are developed constantly, and existing ones are continuously changed and improved. Although most existing languages were written for a specific computer, the newer ones can run on several different systems by design. The language incompatibility between the many different systems may present a problem from a teaching point of view. In practice, because a company normally uses only one such language—regardless of the number of different CNC systems it may operate—this incompatibility does not pose an insurmountable problem.

12.2 SELECTING A COMPUTER-AIDED PROGRAMMING LANGUAGE

In the past few years there have been many advancements in the CAD/CAM and interactive graphics field. These changes were brought about to a large extent by technological advances in the field of computer hardware.

The introduction of Personal Computers (PCs) in 1985 promised to deliver a low-cost CNC programming tool. It became obvious, however, that the Intel 80286 processor used by most PCs did not have the required speed for dealing with state-of-the-art interactive graphics programming. The industry standard MS-DOS operating system did not have the capability to run programs requiring more than 640 KB, due to the lack of virtual memory, nor did it have multiprocessor capability. The programmer could only implement one step at a time (run the program, edit, plot, print, etc.). Time-consuming plotting or printing tied up the system. Initially, networking capabilities were not available. In spite of all these shortcomings, the PCs revolutionized the CNC computer-assisted part programming field.

In 1986 new PCs were enhanced with math coprocessors, high-resolution graphic boards and monitors, drivers, accelerators, and networking. A multitasking

32-bit processor became available. 1988 and 1989 brought definite improvements in the field, by moving toward standards such as X-windows for graphics; OSI— open standard interconnect protocols for networking; and UNIX as the operating system.

The PCs have become the dominant players for CNC programming applications. IBM's OS/2 and Intel's 80386 already provide 4 MIPS (millions of instructions per second) vs. 0.5–1 MIPS for the 286's. Breaking the 640 KB virtual memory limitations opens the barrier to the all-important multitasking option.

Current systems, such as Schlumberger's BRAVO3, accept engineering drawings from another database. The drawing can be used by the CNC programmer to develop the required CNC program for the machining of a specific part. For detailed discussions and examples of BRAVO3, consult the index.

The future part programmer will use emerging artificial intelligence techniques. Many CIM systems are building knowledge bases which can be used in an inference learning and natural language environment. These interactive technologies will bring about a long-awaited user friendly environment.

Some of the fundamentals of computerized part programming will be discussed with practical part programs. Language-based programs used are APT and Compact II, and BRAVO3 was selected as an example of interactive CAD/CAM application.

Note that the BRAVO3 programming has eliminated the language requirements through instant graphic display of geometry and machine motions. Operations are performed by selecting commands from menus and replying to prompts that appear directly above the menus. The source program is a computer record of the part, created by the computer. It contains the part description and machine tool motions required for the making of a specific part.

12.3 THE PROGRAMMING PROCESS

The process of programming consists of logical steps. The programmer must first define the coordinate system in which the CNC program will be written, then define, in terms of computer statements, the part geometry. The type and size of tool as well as the tool motion instructions should be based on the process for achieving the part geometry. This information must be defined in any case.

Using the above information, the CNC computer will perform all the necessary calculations for the postprocessor to code the CNC tape. The postprocessor (link) will prepare a complete tape file including codes required to punch the tape, inch/metric dimensioning identification, ASCII (ISO) or EIA tape coding; motion codes for linear or circular interpolation (G01, G02, G03); canned cycle commands for drilling, boring, tapping, etc.; feeds, speeds, tool changes, and other auxiliary codes for setting the spindle, coolant, clamping, tailstock, etc.

12.4 THE PART GEOMETRY

The objective is to introduce the reader to basic two-dimensional programming, and the definitions will be limited to simple geometry. In order to show the similarity between the various computer languages, APT, COMPACT II, and BRAVO3 have been selected for the subsequent examples.

12.4.1 The Point

The point is the smallest element of geometry, and it is illustrated by examples in both languages selected in Figure 12.2a and b.

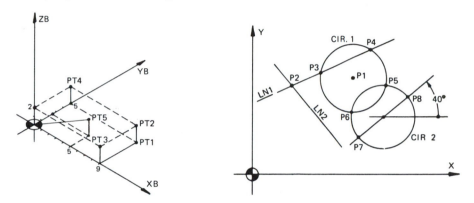

Figure 12.2   Points defined by the intersection of (a) lines and (b) circles.

12.4.1.1 Point defined by coordinates

COMPACT II

```
DPT1,9XB,5YB,ZC
DPT2,9XB,5YB,2ZB
DPT3,9XB,YB,2ZB
DPT4,XB,5YB,2ZB
DPT5,5XB,3YB,2ZB
```

Any point can be defined in terms of the base coordinate system by the following general format:

```
DPTi,aXB,bYB,cZB
```

APT

```
P1 = POINT/9,5,0
P2 = POINT/9,5,2
```

P3 = POINT/9,0,2
P4 = POINT/0,5,2
P5 = POINT/5,3,2

and the general format of the APT definition is:

Pi = POINT/X,Y,Z

12.4.1.2 Point defined by the intersection of lines and circles.   Assuming all the points to be located in the same X-Y plane, the points shown can be defined as follows:

COMPACT II

DPT1,CIR1/CNTR
DPT2,LN1,LN2
DPT3,LN1,CIR1,XS
DPT4,LN1,CIR1,YL
DPT5,CIR1,CIR2,XL
DPT6,CIR1,CIR2,XS
DPT7,CIR2,40CCW
DPT8,CIR2,140CW

The general definitions, corresponding to the Compact II examples, are shown below:

DefinePoinTi,symbol for circle/CeNTeR
DefinePoinTi,symbol for line,symbol for line
DefinePoinTi,symbol for line,symbol for circle,XLarge
                                                     XSmall
                                                     YLarge
                                                     YSmall
DefinePoinTi,symbol for circle,symbol for circle,XLarge
                                                     XSmall
                                                     YLarge
                                                     YSmall
DefinePoinTi,symbol for circle,angle and direction of rotation.

APT

P1 = POINT/CENTER,CIR1
P2 = POINT/INTOF,LN1,LN2
P3 = POINT/XSMALL,INTOF,LIN1,CIR1
P4 = POINT/YLARGE,INTOF,LIN1,CIR1
P5 = POINT/XLARGE,INTOF,CIR1,CIR2
P6 = POINT/XSMALL,INTOF,CIR1,CIR2
P7 = POINT/CIR2,ATANGL,220   or P7 = POINT/CIR2,ATANGL,-140
P8 = POINT/CIR2,ATANGL,40

We are assuming here that the lines and circles used in the definitions of the various points have been defined previously in the same program. The general definitions equivalent to the above specific examples are given in the following APT geometry equations:

```
Pi = POINT/CENTER,symbol for circle
Pi = POINT/INTersectionOF,symbol for first line,symbol for the other line
Pi = POINT/XLARGE
          /XSMALL
          /YLARGE
          /YSMALL,INTersectionOF,symbol for line,symbol for circle
Pi = POINT/XLARGE
          /XSMALL
          /YLARGE
          /YSMALL, INTersectionOF,symbol for first circle, symbol for the other circle
Pi = POINT/Symbol for circle,ATanANGLe,angular value
```

There are many more ways of defining points in both APT and Compact II. Interested readers should consult the geometry sections of the appropriate programming manuals.

12.4.2 The Line

The computer program defines the line as an element of the part geometry. Mathematically, the defined line continues on beyond the part surface, in both directions, to "infinity." Figure 12-3a and b show simple forms of lines most common in the definition of part geometries.

12.4.2.1 Line definitions by coordinates, points, and angles

COMPACT II

```
DLN1,PT1,PT2
DLN2,LN1/1.25YL
DLN22,LN37/0.85XL
DLN2,PT12,PARLN1
DLN3,PT3,PERLN2
DLN4,PT4,PARX/35CW
DLN5,LN4,ROTXY-70 or DLN5,PT5,PARX/35CCW
```

The general definitions are shown below:

```
DefineLiNei,symbol for point,symbol for point
DefineLiNei,symbol for line,dimension parallel,modifier
DLNi,symbol for point,PARallel symbol for line
```

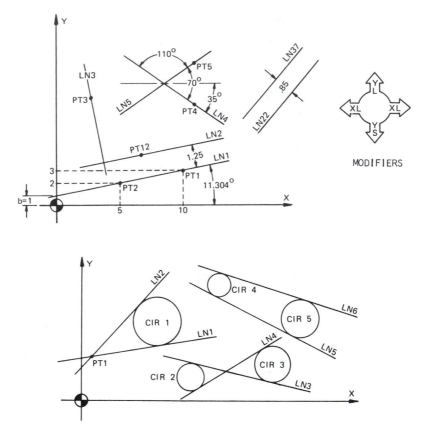

Figure 12.3 Lines defined by (a) coordinates, points, and angles, and (b) using points and circles.

DLNi,symbol for point,PERpendicular symbol for line
DLNi,symbol for point,PARallel symbol for coordinate/dimension and symbol for rotation
DLNi,symbol for line,ROTation symbol for axes with direction and magnitude for rotation.

The same lines are again defined in APT. The wording and format are different; however, the similarity is there.

APT

$$L1 = LINE/5,2,0,10,3,0 \quad \text{or} \quad L1 = LINE/PT1,PT2$$

The line L1 can also be defined using the slope-intercept equation of the line, in its format

$$y = mx + b$$

The slope of the line,

$$m = \frac{y_1 - y_2}{x_1 - x_2} = \frac{3 - 2}{10 - 5} = \frac{1}{5} = 0.2$$

The "y-intercept" is 1.0 inch and line 1 can now be written:

L1 = LINE/SLOPE,0.2,INTERC,1.0

Continuing with the rest of the geometry,

L2 = LINE/PARLEL,L1,YLARGE,1.25
L2 = LINE/P2,ATANGL,0,L1
L2 = LINE/P2,PARLEL,L1

Line 2 (L2) has been defined in three different ways using different modifiers.   In the first case, a parallel condition to a previously defined line (L1) was used, modified in the "Y-large" direction.

In the second case we have used a previously defined line to indicate the direction, by stating a zero (0) angle condition, which is in fact a statement of parallelism.   This definition allows the programmer to call for any angular relationship with the first line, by simply changing the angle zero (0) to whatever angle may be called in the drawing.   P2 defined a fixed point through which our defined line crosses, in the part coordinate system.

In the third case, the word "PARLEL" (i.e., parallel) relates the direction of L1 and the location of P2.

There is no confusion in the selection of symbols when the same line is designated L2 or LN2, for instance.   It is strictly a case of attempting to show two different languages using the same sketch.   The first symbol is preferred in APT-like languages, while the latter is used in Compact II.   Using them interchangeably within the same program would of course lead to an error message.

L3 = LINE/P3,ATANGL,90,L2
L3 = LINE/P3,PERPTO,L2
L4 = LINE/P4,ATANGL,110,L5
L4 = LINE/P4,ATANGL,-70,L5
L5 = LINE/P5,ATANGL,70,L4
L5 = LINE/P5,ATANGL,-110,L4
L22 = LINE/PARLEL,L37,YLARGE,0.85

The programming format for the L22 statement is:

Li = LINE/PARLEL, symbol for line,XLARGE
                                  XSMALL
                                  YLARGE
                                  YSMALL, magnitude of offset.

12.4.2.2 Line definitions by previously defined circles and points

COMPACT II

DLN1,PT1,CIR1,YS
DLN2,PT1,CIR1,YL
DLN3,CIR2,YL,CIR3,CROSS
DLN4,CIR2,YS,CIR3,CROSS
DLN5,CIR4,YS,CIR5
DLN6,CIR4,YL,CIR5

APT

L1 = LINE/P1,RIGHT,TANTO,C1
L2 = LINE/P1,LEFT,TANTO,C1
L3 = LINE/LEFT,TANTO,C3,RIGHT,TANTO,C2
L4 = LINE/LEFT,TANTO,C3,RIGHT,TANTO,C2
L5 = LINE/RIGHT,TANTO,C4,RIGHT,TANTO,C5
L6 = LINE/LEFT,TANTO,C4,LEFT,TANTO,C5

12.4.3 The Circle

The circle is the third most common geometric shape after the point and the line. In the geometry section, the programmer defines the circle in its entirety. In the tool motion section, the program outlines which portion of this circle will be machined as part of the component boundary. Figure 12.4 shows some of the more common definitions.

COMPACT II

DCIR1,P1,0.5R
DCIR1,PT2,PT3,PT4

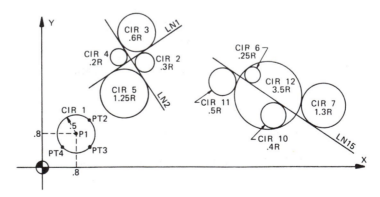

Figure 12.4 Circle definitions.

Circle 1 is first defined through the location of its center and the magnitude of its radius, while in the second statement it is defined through three points on its circumference.

DCIR2,LN1/0.3YS,LN2/0.3XL,0.3R

This rather involved statement defines the circle as follows: The center is located at the intersection of two lines, one offset by 0.3 inch below line 1 and the other offset by the same amount to the right of line 2. The value of the circle radius is eventually defined as 0.3 inch.

DCIR3,LN1/0.6YL,LN2/0.6XL,0.6R
DCIR4,LN1/0.2XS,LN2/0.2YS,0.2R
DCIR5,LN1/1.25YS,LN2/1.25XS,1.25R
DCIR7,LN15/1.3YL,CIR12/1.3XL,1.3R
DCIR11,LN15/0.5YS,CIR12/0.5XS,0.5R

APT

As mentioned before, the APT program will accept most labels for a given geometry. CIR1 or C1 will readily remind the programmer that circle 1 was defined, but so long as a label selected is not part of a list of "forbidden" terms, used by the processor itself and therefore potentially leading to confusion, any symbol may be selected. Symbols must be defined before they can be used in statements.

CIR1 = CIRCLE/0.8,0.8,0,0.5
CIR1 = CIRCLE/CENTER,P1,RADIUS,0.5

In the first definition, the first three coordinates locate the center of the circle in X, Y, and Z, while the last dimension is the circle radius. The second definition achieves the same objective in a more obvious fashion, using the predefined label of point 1 where the center of the circle is located.

CIR2 = CIRCLE/YLARGE,L2,XLARGE,L1,RADIUS,0.3
CIR3 = CIRCLE/XLARGE,L2,YLARGE,L1,RADIUS,0.6
CIR4 = CIRCLE/XSMALL,L2,YLARGE,L1,RADIUS,0.2
CIR5 = CIRCLE/XSMALL,L2,YSMALL,L1,RADIUS,1.25
CIR7 = CIRCLE/YLARGE,L15,XLARGE,OUT,CIR12,RADIUS,1.3
CIR6 = CIRCLE/YLARGE,L15,XSMALL,IN,CIR12,RADIUS,0.25
CIR11 = CIRCLE/YSMALL,L15,XSMALL,OUT,CIR12,RADIUS,0.5
CIR10 = CIRCLE/YSMALL,L15,YSMALL,IN,CIR12,RADIUS,0.4

In both the APT and Compact II languages, there are a great deal more definitions for points, lines, and circles. In addition, more complex geometric solutions are available for vectors, planes, cones, ellipses, and general quadratic surfaces which include hyperbolic paraboloids, ellipsoids, hyperboloids, etc.

It should also be mentioned that most "computer" words may be shortened to a single letter if required, thus decreasing the length of the program and the time required to write it.

Further in-depth discussion on the topics of geometry definition are beyond the objectives of this book. Interested readers are advised to obtain the programming manual corresponding to the language of their choice.

12.5 TOOL OR CUTTER STATEMENTS

Tool or cutter statements must be defined prior to any motion statements. The part geometry, described in the geometry definition section, is written using the dimensions of the component. Any tool motion without having specified a tool diameter or radius would create a tool path for a zero diameter tool. Specifying a cutter diameter will alter the tool path achieving cutter diameter compensation.

The tool statements are discussed in more detail in the sample parts described in this chapter, as there are significant differences between milling and turning programming applications.

12.6 MOTION STATEMENTS

Motion statements are often called "machine control instructions." All CNC computer-assisted languages provide the programmer with a variety of statement formats for describing tool or machine motions, usually limited to the part geometry specified in the program. Other machine functions associated with tool movements, such as acceleration, deceleration, dwell, home, etc., are often interactive with a specific postprocessor or link. It also has to be pointed out that formats and wording statements cannot be changed and are to be used exactly as prescribed in the respective programming manuals. Manuscripts should be checked for accuracy and validity prior to entering the program into the computer. Most programs can be corrected off-line. On-line corrections should only be performed by experienced programmers.

12.7 COMPACT II MILLING—SAMPLE PROGRAMS

In the following pages we present a number of sample programs for milling and turning applications, including explanations wherever required for clarification. These sample programs were compiled on a Compact II in-house system, including a Hewlett-Packard plotter. The sample parts were machined on an OKK 410 machining center with a Fanuc 6MB control, and a 4NE 12-position rear turret turning center with a Fanuc 6T control.

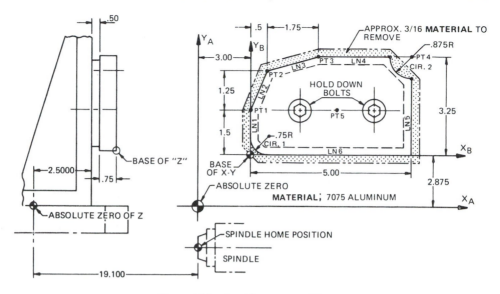

Figure 12.5   Outside contour milling.

12.7.1 Example

Write a Compact II part program for the outside contour milling of the sample part illustrated in Figure 12-5. The program should include both rough and finish milling. Specific information must be available to the programmer prior to the start of the program. This includes cutter position at start, cutter diameter and length, setup data, including the name of the link (postprocessor). The following is a detailed listing.

12.7.1.1 The name of the link.   The name of the link must be listed as the first statement of the program.

MACHIN,SEIKIFAN2

The name "SEIKIFAN2" represents a specific machine and control combination.

12.7.1.2 Identification.   Identification is the second statement of the program, and its contents will be punched out in tape in readable hole patterns in front of the machine control tape.

IDENT,SAMPLE PART,PARTNO.12345

12.7.1.3 Initialization.   Initialization is the third statement in the program, and it will set the input and output modes for dimensional computation.

INIT,INCH/IN,INCH/OUT

The format shown above indicates that the part dimensions and tool sizes and motions will be given in inches, as will be the motions output in the machine control tape. Other possibilities of the above statement are:

INIT,INCH/IN,METRIC/OUT
INIT,METRIC/IN,INCH/OUT and
INIT, METRIC/IN,METRIC/OUT

Only one initialization may be used in the CNC program.

12.7.1.4 Setup.   *Setup* is the fourth statement in the program. It is used to establish travel limits for the tool, specify the home position (where the tool changes will take place, normally) and the machine absolute zero. Additional information may be included in the setup to describe special link requirements for specific systems.

SETUP,CMOD/1/2,MOD500/2,ABSO1,PALLET,LX,LY,12.50LZ,RPM3150

The first two statements after the word "SETUP" are link-related identifiers of specific equipment within a family of systems.

ABSO1 instructs the CNC computer to calculate all tool motions in absolute mode. It will also generate a tape block to "zero" the control at its absolute zero position:

N G92 X0 Y0 Z0

This statement will cause the control positional registers to show zero. As the machine will have been "zeroed" prior to running the tape, what takes place may be described as a synchronization of the machine and the control.

PALLET will generate a tape block for pallet transfer prior to any tool motion that takes place. This machine has two pallets, one for machining, the other for loading-unloading. The corresponding tape block is:

N M60

nLX represents the distance along the axis from absolute zero (i.e., machine origin) to home position. As in our example $n = 0$, there is no distance between the above locations, and tool change will take place at machine origin. nLY is the distance along the Y-axis from absolute zero to home position. In our case, this $n = 0$ as well.

nLZ is the distance along the Z-axis from absolute zero to home position. In our example, this dimension is 12.50 inches.

RPMN is the spindle speed word, limiting the maximum programmable spindle speed at 3150 rpm. It also indicates to the control that on this system, the machine rpm may be programmed directly, without requiring M codes for various gear ranges.

12.7.1.5 Base.   Base is the fifth statement in the program. The function of the base statement is to establish a base or coordinate system on the part for easier programming. In many cases, it coincides with the drawing datum point. The base is also used as a start point in the program.

In our example, the base is established at the front-left-top corner of the part, using the statement:

BASE,3XA,2.875YA,3.75ZA

The base has now been located along the three axes from the machine origin ("machine absolute zero"). We can now proceed with the problem solution in the following section.

12.7.1.6 Solution

We have seen the start-up portion of the program. We are relisting it below:

```
MACHIN,SEIKIFAN2
IDENT,SAMPLE,PART,PARTNO.12345
INIT,INCH/IN,INCH/OUT
SETUP,CMOD5/1/2,MOD500/2,ABSO1,PALLET,LX,LY,12.50LZ,RPM3150
BASE,3XA,2.875YA,3.75ZA
```

We will now establish and write the part geometry statements.

Points

```
DPT1,XB,1.5YB,ZB
DPT2,0.5XB,2.75YB,ZB
DPT3,2.25XB,3.25YB,ZB
DPT4,5.0XB,3.25YB,ZB
DPT5,2.5XB,1.625YB,ZB
```

Every time a point is written in the program, it should also be labeled on the programmer's drawing. This is required for subsequent ease of clarification, program checking, and debugging. Point 5 will be used to locate the part plot on the plotter, and it will be used in conjunction with a "draw" statement.

The reader has noticed that all the above points have been defined in the part coordinate system (XB, YB, and ZB). The Compact II language, however, allows the programmer to use the machine coordinate system as well, or any combination of the two, or another base if correctly defined. To illustrate this important capability, the definition of point 1 will be rewritten below using all possible combinations allowed so far by the system:

```
DPT1,3XA,4.375YA,3.75ZA
DPT1,3XA,1.5YB,ZB
DPT1,3XA,4.375YA,ZB
DPT1,XB,1.5YB,3.75ZA
DPT1,XB,4.375YA,3.75ZA
```

Obviously the programmer should only use one of these statements in the program. The reader should not, however, in spite of the freedom of the language, take advantage of mixed mode capabilities. This complicates the program and makes corrections more difficult. A change in the base statement would also require the programmer to change every statement that contains mixed mode definitions.

Lines

```
DLN1,XB
DLN2,PT1,PT2
DLN3,PT2,PT3
DLN4,3.25YB
DLN5,5XB
DLN6,YB
```

Circles

```
DCIR1,LN6/0.75YL,LN1/0.75XL,0.75R
DCIR2,PT4,0.875R
```

Having completed the geometry section, we shall now program a "draw" statement for the plotting of the part geometry and tool path.

```
DRAW, SCALE1, PT5,CNTR
```

The scale statement allows the programmer to increase or decrease the size of the plot in relation to the dimensions of the drawing. For example,

SCALE0.5 will result in a half-size plot
SCALE2 will provide us with a plot twice the real part size.
PT5,CNTR instructs the computer to locate the center of the plot at point 5.

Tool change

Tool change contains all the information relevant to machining and as such must precede the first motion statement in the program. Machines equipped with an automatic tool changer (ATC) are programmed using the word ATCHG. This must be programmed every time a tool change is required. In addition, this word initiates a number of other events, such as the axes being moved to the tool change point, the spindle and coolant being turned off, the spindle being oriented, the tool magazine being indexed, the spindle speed being selected, the spindle being started, the feed rate being set, etc.

ATCHG,TOOL1,5.75GL,0.500TD,800RPM,8IPM,NOX,NOY,CON,0.05STK

where

> TOOL1 will generate the appropriate code for the tool change. Machines equipped with random tool changers will index to position 1 and change, while sequential tool changers will select the next available tool in the turret as tool No. 1.
>
> nGL specifies the preset length of the tool. As shown in Figure 12.6, $n =$ 5.75 inches.
>
> nTD is the tool diameter, necessary to the program for calculating the tool path. In our program $n = 0.5$ inches.
>
> nRPM specifies the spindle speed in revolutions per minute. It will generate the following tape block:

N S800 M03

> nIPM indicates the feed rate. In the program, $n = 8$ ipm.
>
> NOX,NOY normally specify that no X-motion and no Y-motion are to take place, i.e., the tool is not to go "home" to its prescribed tool change location. As in our case the programmed tool change takes place at machine absolute zero, these words are not really required; they are in strictly for training purposes.
>
> CON, meaning coolant on, will generate a tape block to turn the coolant on after the tool change has been completed.
>
> nSTK specifies the thickness of material to be left for finishing, as "n" inches, in our case, 0.05. This amount will be added by the program to the cutter radius for compensation.

There are additional features that can be programmed in the tool change

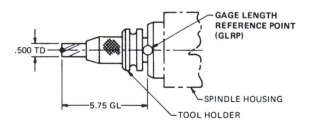

GAGE LENGTH
REFERENCE POINT
(GLRP)

.500 TD

5.75 GL

SPINDLE HOUSING

TOOL HOLDER

Figure 12.6 Tool gage length.

instructions for specific purposes. As for the rest of the details, this information is available in the specific programming manual.

Motion statements

The programmer must use previously defined part geometry to specify the desired motion pattern for the machine. The motion statements are many and varied; therefore, specific statements will be explained as used in the program. For our part, two passes have been specified around the part surface. The roughing pass will leave 0.050 inch stock on the outside contour, which will be removed by the finishing pass. To avoid writing the program twice, we have the capability of numbering the first and last motion statement in the machining sequence. Once the roughing cut has been programmed, the finishing can be programmed by only using a "Do" statement as shown below.

$\langle 1 \rangle$MOVE, TOLN1/XS, TOLN6/0.3YS,0.1ZB

This statement will move the tool to point A, as illustrated in Figure 12.7.

The word "MOVE" generates rapid tool motion, equivalent to the manual instruction

N . . . . . G00 Y . . . Y . . . Z . . .

$\langle 1 \rangle$ represents statement number 1.

CUT,0.8ZB

This instruction will move the tool to point B.

The word "CUT" will generate tool motion in controlled feed rate. The amount of the feed rate was programmed in the "ATCHG" statement as 8 ipm.

CUT,PARLN1,PASTLN2

This statement, read as "cut, parallel to line 1 until the cutter gets completely on

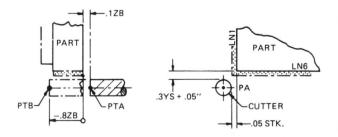

Figure 12.7   Clearance and stock.

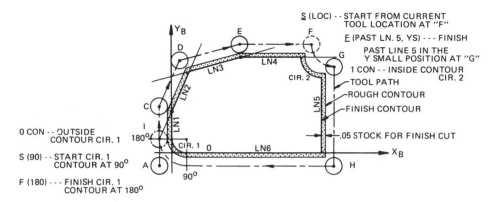

Figure 12.8   Programmers sketch for outside contour milling.

the other side of line 2" will bring the tool to point C, as shown in Figure 12.8

CUT, PARLN2, PASTLN3	Tool to point D
CUT, PARLN3, PASTLN4	Tool to point E
CUT, PARLN4, INCIR2, XS	Tool to point F
ICON, CIR2, CCW, S(LOC), F(PASTLN5,YS), 0.875R	Tool to point G
CUT, PARLN5, PASTLN6	Tool to point H
⟨2⟩OCON, CIR1, CW, S(90),F(180)	Tool to point I
ATCHG, TOOL2, 6.0GL, 0.5TD, 140 FPM, 0.005IPR, CON,NOX,NOY	

In the second tool change, above, different words were used to specify tool feed and speed.

nFPM specifies the tool surface velocity as $n = 140$ feet per minute. The CNC computer will calculate the required spindle (tool) rpm using the equation

$$n = 12 \cdot v/\pi \cdot d$$

nIPR specifies the cutting feed as $n = 0.005$ inches per tool revolution.

S(LOC) means start from the location of the tool at the time the statement is read, on our part at point F.

F(PASTLN5,YS) indicates that the finish of the circular interpolation motion will see the cutter past line 5. There are two possibilities for this definition, after a 90° and a 270° angle. The first position is at G, below the latter, on the Y-axis, hence YSmall.

ICON is short for inside contour, in our case, that of circle 2.

OCON represents the outside contour of circle 1.

S(90), respectively, F(180) represent a start of the contour at 90° and an end at 180°, measured clockwise from the positive X-axis which is always 0° in Compact II.

DO1/2,O STK

This DO statement instructs the CNC computer to establish a new tool path for the finishing cut. The instructions located in the program between motion statements ⟨1⟩ and ⟨2⟩ will be carried out, using the tool data specified in the "ATCHG" statement for tool No. 2.

END will terminate the program by generating an end of program tape block, as N M30

Processing the Compact II Program

This program, once entered into the disk file, was processed by the Compact II processor in the substitute run command mode. During the interactive process, the computer was instructed to create a "list file" and a "tape file."

The list file, along with the tool path and part geometry plots, as shown in Figure 12.9, allows the programmer to check the program output and make changes as required. Once the programmer is satisfied that the program is correct, the computer is instructed to punch the machine control tape (off the tape file) and print its content. As instructed by the programmer, the machine control tape will be punched out in EIA or ASCII (ISO) format.

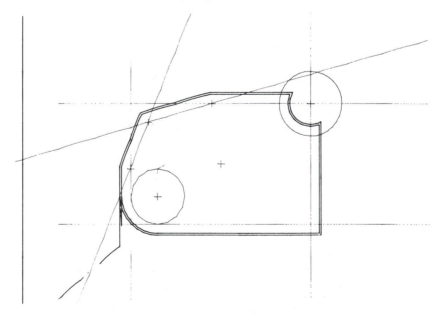

Figure 12.9 Computer-produced plot for outside contour milling.

Source program

```
+ P
PUNCHING TAPE TOO?
IDENT?
MACHINE,SEIKIFAN2
IDENT,SAMPLE PART,PART NO. 12345
INIT,INCH/IN, INCH/OUT
SETUP,CMOD5/1/2,MOD500/2,ABSO1,PALLET,LX,LY,LZ,RPM3150
BASE,3XA,2.875YA,3.75ZA
DPT1,XB,1.5YB,ZB
DPT2,.5XB,2.75YB,ZB
DPT3,2.25XB,3.25YB,ZB
DPT4,5.0XB,3.25YB,ZB
DPT5,2.5XB,1.625YB,ZB
DLN1,XB
DLN2,PT1,PT2
DLN3,PT2,PT3
DLN4,3.25YB
DLN5,5.0XB
DLN6,YB
DCIRI,LN6/.75YL,LN1/.75XL,.75R
DCIR2,PT4,.875R
DRAW,SCALE1,PT5,CNTR
ATCHG,TOOL1,5.75GL,.500TD,800RPM,8IPM,NOX,NOY,CON,.05STK
<1>MOVE,TOLN1/XS,TOLN6/.3YS,.1ZB
CUT,.8Z
CUT,PARLN1,PASTLN2
CUT,PARLN2,PASTLN3
CUT,PARLN3,PA3TLN4
CUT,PARLN4,INCIR2,XS
ICON,CIR2,CCW,S(LOC),F(PASTLN5,YS),.875R
CUT,PARLN5,PASTLN6
<2>OCON,CIR1,CW,S(90),F(180)
ATCHG,TOOL2,6.0GL,.5TD,140FPM,.005IPR,CON,NOX,NOY
DO1/2,OSTK

END
```

List file

```
> MACHINE,SEIKIFAN2
MAIN 60980 LINK 21580 SYS 42578 L# 1483

> IDENT,SAMPLE PART,PART NO. 12345
00-00-00 00:00

> INIT,INCH/IN,INCH/OUT
```

```
> SETUP,CMOD5/1/2,MOD500/2,ABSO1,PALLET,LX,LY,LZ,RPM3150

> BASE,3XA,2.875YA,3.75ZA
= X 3. Y2.875 Z 3.75

> DPT1,XB,1.5YB,ZB
= X 3. Y 4.375 Z 3.75

> DPT2,.5XB,2.75YB,ZB
= X 3.5 Y 5.625 Z 3.75

> DPT3,2.25XB,3.25YB,ZB
= X 5.25 Y 6.125 Z 3.75

> DPT4,5.0XB,3.25YB,ZB
= X 8. Y 6.125 Z 3.75

> DPT5,2.5XB,1.625YB,ZB
= X 5.5 Y 4.5 Z 3.75

> DLN1,XB
= X 3. Y . A 90.

> DLN2,PT1,PT2
= X 3. Y 4.375 A 68.1986

> DLN3,PT2,PT3
= X 3.5 Y 5.625 A 15.9454

> DLN4,3.25YB
= X . Y 6.125 A .

> DLN5,5.0XB
= X 8. Y. A 90.

> DLN6,YB
= X . Y 2.875 A .

> DCIR1,LN6/.75YL,LN1/.75XL,.75R
= X 3.75Y 3.625 R .75

> DCIR2,PT4,.875R
= X 8. Y 6.125 R .875

> DRAW,SCALE1,PT5,CNTR

> ATCHG,TOOL1,5.75GL,.500TD,800RPM,8IPM,NOX,NOY,CON,.05STK
  N0000 M60
  N0010 G92 X0 Y0 Z0
  N0020 G91 G28 Z0
  N0030 G90 G00 X0 Y0
  N0040 M06
  N0050 S800 M03

> <1>MOVE,TOLN1/XS,TOLN6/.3YS,.1ZB
  N0060 G90 G00 Z-9.6000 M08
```

```
    N0070 X 2.7000 Y 2.2750

> CUT,.8Z
    N0080 G01 Z-10.4000 F8.00

> CUT,PARLN1,PASTLN2
    N0090 Y 4.4328

> CUT,PARLN2,PASTLN3
    N0100 X 3.2761 Y5.8730

> CUT,PARLN3,PASTLN4
    N0110X 5.2081 Y 6.4250

> CUT,PARLN4,INCIR2,XS
    N0120 X 7.5095

> ICON,CIR2,CCW,S(LOC),F(PASTLN5,YS),.875R
    N0130 G03 X 7.4250 Y 6.1250 I .4905 J-.3000 F5.58
    N0140 X 8.0000 Y 5.5500 I .5750
    N0150 X 8.3000 Y 5.6345 J .5750

> CUT,PARLN5,PASTLN6
    N0160 G01 Y 2.5750 F8.00

> <2>OCON,CIR1,CW,S(90),F(180)
    N0170 X 3.7500
    N0180 G02 X 2.7000 Y3.6250 J 1.0500

> ATCHG,TOOL2,6.0GL,.5TD,140FPM,.005IPR,CON,NOX,NOY
    N0190 G00
    N0200 Z0
    N0210 M06

> DO1/2,0STK
*  D0
    N0220 S1069 M03
    N0230 G90 G00 Z-9.8500 M08
    N0240 X 2.7500 Y2.8250

*  DO
    N0250 G01 Z-10.6500 F5.35
*  DO
    N0260 Y 4.4232
*  DO
    N0270 X 3.3134 Y 5.8317
*  DO
    N0280 X 5.2150 Y 6.3750
*  DO
    N0290 X 7.4272
*  DO
    N0300 G03 X 7.3750 Y 6.1250 I .5728 J-.2500 F3.82
    N0310 X 8.0000 Y 5.5000 I .6250
```

```
    N0320 X 8.2500 Y 5.5522 J .6250
*  DO
    N0330 G01 Y 2.6250 F5.35
*  DO
    N0340 X 3.7500
    N0350 GO2 X 2.7500 Y 3.6250 J 1.0000

>

> END
    N0360 G00 X0 Y0
    N0370 ZO
    N0380 M30
END MIN: 6.2 FT: 14.5 MTR: 4.4

ERRORS DETECTED = 0
```

Tape file

```
TAPE FILE: /CNC-JP100T/
EIA?
NO
OUTPUT TO: TPT
TURN PUNCH ON,HIT CR
N0000M60
N0010G92X0Y0Z0
N0020G91G28Z0
N0030G90G00X0Y0
N0040M06
N0050S800M03
N0060G90G00Z-96000M08
N0070X27000Y22750
N0080G01Z-104000F800
N0090Y44328
N0100X32761Y58730
N0110X52081Y64250
N0120X75095
N0130G03X74250Y61250I4905J-3000F558
N0140X80000Y55500I5750
N0150X83000Y56345J5750
N0160G01Y25750F800
N0170X37500
N0180G02X27000Y36250J10500
N0190G00
N0200Z0
N0210M06
N0220S1069M03
N0230G90G00Z98500M08
N0240X27500Y28250
```

N0250G01Z-106500F535
N0260Y44232
N0270X33134Y58317
N0280X52150Y63750
N0290X74272
N0300G03X73750Y61250I5728J-2500F382
N0310X80000Y55000I6250
N0320X82500Y55522J6250
N0330G01Y26250F535
N0340X37500
N0350G02X27500Y36250J10000
N0360G28X0Y0
N0370G28Z0
N0380M30

12.7.2 Example

Write a Compact II program for outside contour milling, hole pattern drilling and tapping, for the part illustrated in Figure 12.10. The program should leave 0.1 inch stock on the outside contour for finish milling.

PROGRAMMING DATA

Tool No. 1: 0.5-inch diameter, 6.1 GL (gage length), end mill
Tool No. 2: 0.125-inch diameter, 5.5 GL, center drill
Tool No. 3: 0.3125-inch diameter 5.75 GL, drill
Tool No. 4: 0.375-inch diameter 5.25 GL, tap

The rest of the machine and link data will be identical to the previous example.

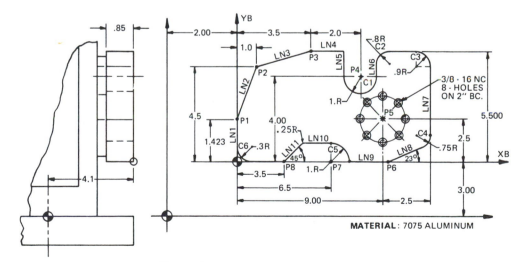

Figure 12.10 Milling, drilling, and tapping.

12.7.2.1 Solution

Start-up

MACHIN, SEIKIFAN2
IDENT, SAMPLE PART MILL-DRILL-TAP
INIT,INCH/IN, INCH/OUT
SETUP,CMOD/1/2,MOD500/2,ABSO1,PALLET,XL,YL,12.50LZ,RPM3150
BASE,2XA,3YA,4.1ZA

Geometry

DPT1,XB,1.423YB,ZB
DPT2,1.0XB,4.5YB,ZB
DLN1,XB
DLN2,PT1,PT2
DPT3,3.5XB,5.5YB,ZB
DLN3,PT2,PT3
DLN4,5.5YB
DLN5,4.5XB
DLN6,6.5XB
DLN7,11.5XB
DPT4,5.5XB,4YB,ZB
DCIR1,PT4,1.0R
DPT5,9.0XB,2.5YB,ZB
DPT6,9.0XB,YB,ZB
DLN8,PT6,23CCW
DLN9,YB
DLN10,1.0YB
DPT7,6.5XB,YB,ZB
DPT8,3.5XB,YB,ZB
DLN11,PT8,45CCW
DCIR2,LN6/0.8XL,LN4/0.8YS,0.8R
DCIR3,LN4/0.9YS,LN7/0.9XS,0.9R
DCIR4,LN7/0.75XS,LN8/0.75YL,0.75R
DCIR5,PT7,1.0R
DCIR6,LN9/0.3YL,LN1/0.3XL,0.3R

Plot

DRAW,SCALE (1/2), PT5,CNTR

First tool change for milling

ATCHG,TOOL1,6.1GL,0.5TD,850RPM,8.5IPM,0.1STK,CON

Rough milling

```
⟨4⟩MOVE,OFFLN1/XS,OFFLN9/0.3YS,0.1ZB
CUT,0.9ZB
CUT,PARLN1,OFFLN2/YL
CUT,PARLN2,PASTLN3
CUT,PARLN3,PASTLN4
CUT,PARLN4,PASTLN5
ICON,CIR1,CCW,S(180),F(0)
OCON,CIR2,CW,S(180),F(270)
OCON,CIR3,CW,S(270),F(0)
OCON,CIR4,CW,S(0),F(TANLN8)
CUT,PARLN8,OFFLN9/YS
CUT,PARLN9,INCIR5,XL
ICON,CIR5,CCW,S(0),F(90)
CUT,PARLN10,OFFLN11//XL
CUT,PARLN11,OFFLN9/YS
⟨5⟩OCON,CIR6,CW,S(90),F(180)
DO 4/5,0 STK
```

Interpretation of motion commands OFFLNi/XL and INsideCIRclei,XL

/XS	XS
/YL	YL
/YS	YS

Using Figure 12.11,
from point A to point B: CUT,PARLN1,OFFLN2/XS
from point A to point C: CUT,PARLN1,OFFLN2/XL
from point C to point D: CUT,PARLN2,OFFLN3/YL
from point C to point E: CUT,PARLN2,OFFLN3/YS
from point E to point F: CUT,PARLN3,OUTCIR1,XL
from point E to point G: CUT,PARLN3,INCIR1,XL
from point E to point H: CUT,PARLN3,INCIR1,XS
from point E to point K: CUT,PARLN3,OUTCIR1,XS

Center drilling

```
ATCHG,TOOL2,5.5GL,0.125TD,2100RPM,4IPM,NOX,NOY,CON,118TPA
DRL,PT5,2.0BC,8EQSP,CW,S(0),0.3DP
```

where

DRL is a major word that initiates the drill cycle
PTi specifies the number of the point where the bolt circle center is located
iEQSP represents the number of equally spaced holes, in our case i = 8

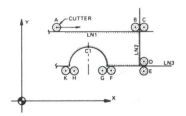

Figure 12.11   Sketch illustrating various motion commands.

CW specifies the direction, clockwise, in which the holes will be successively drilled

S(i) is the starting angle, respectively the starting angular location of the first hole. In the program, i = 0 on the bolt circle

nBC represents the diameter n = 2 of the bolt circle

iDP specifies the depth of drilling from 0ZB, in our case, i = 0.3

iTPA specifies the tool point angle i = 118°

Drilling

```
ATCHG,TOOL3,5.75GL,0.3125TD,680RPM,5IPM,NOX,NOY,CON,118TPA
DRL,PT5,2.0BC,8EQSP,CW,S(0),0.85THRU
```

iTHRU specifies that the holes are to be drilled through the part whose thickness i = 0.85 inches.

Tapping

```
ATCHG,TOOL4,5.25GL,0.375TD,(1/16)LEAD,100RPM,NOX,NOY,CON
FLT,PT5,2.0BC,8EQSP,CW,S(0),0.85THRU
```

iLEAD represents the lead or pitch of the tapped hole; i = 1/16 = 0.0625 inch. The Compact II computer will use this value, in conjunction with the rpm to compute the tap feed.

END

The list and tape file printouts are shown below, followed by Figure 12.12, which illustrates the part geometry and tool path plot for the sample part.

12.7.2.2 Source program

```
MACHINE,SEIKIFAN2
IDENT,MILL-DRILL-TAP
INIT,INCH/IN,INCH/OUT
SETUP,CMOD5/1/2,MOD500/2,ABSO1.,PALLET,LX,LY,LZ,RPM3150
```

```
BASE,2XA,3YA,4.1ZA
DPT1,XB,1.423YB,ZB
DPT2,1.0XB,4.5YB,ZB
DLN1,XB
DLN2,PT1,PT2
DPT3,3.5XB,5.5YB,ZB
DLN3,PT2,PT3
DLN4,5.5YB
DLN5,4.5XB
DLN6,6.5XB
DLN7,11.5XB
DPT4,5.5XB,4YB,ZB
DPT5,9XB,2.5YB,ZB
DPT6,9.7XB,YB,ZB
DLN8,PT6,23CCW
DLN9,YB
DLN10,1YB
DCIR1,PT4,1R
DPT7,6.5XB,YB,ZB
DPT8,3.5XB,YB,ZB
DLN11,PT8,45CCW
DCIR2,LN6/.8XL,LN4/.8YS,.8R
DCIR3,LN4/.9YS,LN7/.9XS,.9R
DCIR4,LN7/.75XS,LN8/.75YL,.75R
DCIR5,PT7,1.0R
DCIR6,LN9/.3YL,LN1/.3XL,.3R
DRAW,SCALE(1/2),PT5,CNTR
ATCHG,TOOL1,6.1GL,.5TD,850RPM,8.5IPM,.1STK,CON
<1>MOVE,OFFLN1/XS,OFFLN9/.3YS,.1ZB
CUT,.9ZB
CUT,PARLN1,OFFLN2/YL
CUT,PARLN2,PASTLN3
CUT,PARLN3,PASTLN4
CUT,PARLN4,PASTLN5
ICON,CIR1,CCW,S(180),F(0)
OCON,CIR2,CW,S(180),F(270)
OCON,CIR3,CW,S(270),F(0)
OCON,CIR4,CW,S(0),F(TANLN8)
CUT,PARLN8,OFFLN9/YS
CUT,PARLN9,INCIR5,XL
ICON,CIR5,CCW,S(0),F(90)
CUT,PARLN10,OFFLN11/XL
CUT,PARLN11,OFFLN9/YS
<2>OCON,CIR6,CW,S(90),F(180)
ATCHG,TOOL2,5.5GL,.125TD,2100RPM,4IPM,NOX,NOY,CON,118TPA
DRL,PT5,2.0BC,8EQSP,CW,S(0),.3DP
ATCHG,TOOL3,5.75GL,.3125TD,680RPM,5IPM,NOX,NOY,CON,118TPA
DRL,PT5,2.0BC,8EQSP,CW,S(0),.85THRU
```

ATCHG,TOOL4,5.25GL,.375TD,(1/16)LEAD,100RPM,NOX,NOY,CON
FLT,PT5,2.BC,8EQSP,CW,S(0),.85THRU
END

+

12.7.2.3 List file

>MACHINE,SEIKIFAN2
MAIN 60980 LINK 21580 SYS 42578 L# 1483

>IDENT,MILL-DRILL-TAP
00-00-00 00:00

>INIT,INCH/IN,INCH/OUT

>SETUP,CMOD5/1/2,MOD500/2,ABSO1.,PALLET,LX,LY,LZ,RPM3150

>BASE,2XA,3YA,4.1ZA
= X 2. Y 3. Z 4.1

>DPT1,XB,1.423YB,ZB
= X 2. Y 4.423 Z 4.1

(NOTE: The rest of the list file was omitted. Its sole purpose is to relate the instruction and its outcome. The latter is more usefully reflected by the tape file.)

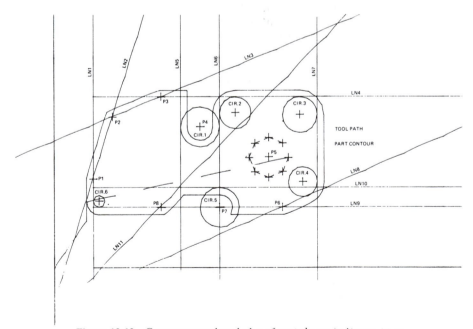

Figure 12.12   Computer-produced plot of part shown in Figure 12.10.

12.7.2.4 Tape file

```
P
& +        SEIKIFAN2           00-00-00 00:00
MILL-DRILL-TAP
=
N0000M60
N0010G92X0Y0Z0
N0020G91G28Z0
N0030G90G00X0Y0
N0040M06
N0050S850M03
N0060G90G00Z-103000M08
N0070X16500Y23500
N0080G01Z-111000F850
N0090Y44785
N0100X27178Y77641
N0110X54326Y88500
N0120X68500
N0130Y70000
N0140G03X75000Y63500I6500F614
N0150X81500Y70000J6500
N0160G01Y77000F850
N0170G02X93000Y88500I11500
N0180G01X126000
N0190G02X138500Y76000J-12500
N0200G01Y42605
N0210G02X131798Y32479I-11000
N022G01X117712Y26500
N0230X90477
N0240X91500Y30000
N0250G03X85000Y36500I-6500F614
N0260G01X66450F850
N0270X5640Y26500
N0280X23000
N0290G02X16500Y33000J6500
N0300G00
N0310Z0
N0320M06
N0330S2100M03
N0350G81X120000YS5000Z-92624R-97000F400
N0360X117071Y47929
N0370X110000Y45000
N0380X102929Y47929
N0390X100000Y55000
N0400X102929Y62071
N0410X110000Y65000
N0420X117071Y62071
```

```
N0430G80Z0
N0440M06
N0450S680M03
N0470G81X120000Y55000Z-88061R-99500F500
N0480X117071Y47929
N0490X110000Y45000
N0500X102929Y47929
N0510X100000Y55000
N0520X102929Y62071
N0530X110000Y65000
N0540X117071Y62071
N0550G80Z0
N0560M06
N0570S100M03
N0590G84X120000Y55000Z-84000R-94500F625
N0600X117071Y47929
N0610X110000Y45000
N0620X102929Y47929
N0630X100000Y55000
N0640X102929Y62071
N0650X110000Y65000
N0660X117071Y62071
N0670G80M03
N0680G28X0Y0
N0690G28Z0
N0700M30
*
```

The plot allows the programmer to check the part geometry (points, lines, circles, etc.) against the tool path with ease. Most simple plots use two colors, one for the geometry and the other for the tool path.

Plots of complex parts using numerous tools can be color-coded with a different color pen for each path. The CNC computer, based on the program information, has completed all the necessary calculations for the tool path and has printed the information required in the list file. The link (postprocessor) received this calculated data and translated it into acceptable codes for the respective CNC machine, including the appropriate circular interpolation codes (G02 and G03) with the correct dimensional words (*X* and *Y*) and unit vectors (*I* and *J*). It selected the appropriate codes and formats for drilling and tapping (G81 and G84). The reader can clearly appreciate that any comparison between manual and computer-assisted programming is weighed in favor of the computer.

12.8 COMPACT II TURNING—SAMPLE PROGRAM

This chapter on computer-assisted programming would not be complete without a turning sample program. The basic definitions are similar, in many cases identical,

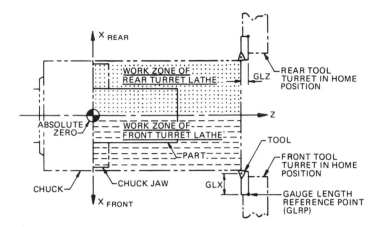

Figure 12.13   Coordinate system of turning center.

to milling. However, there are some essential differences, derived from the concept of machining itself. In turning, the part is rotated, and the tool moves in a two-dimensional plane to generate the prescribed contour. The motion statements are far simpler to analyze in the absence of a third axis.

The most popular CNC is the horizontal programmable tool changer type, best known in industry as the CNC turning center. These are built with "front" or "rear" turrets (the four-axis turning centers have both front and rear tool turrets, independently programmable for simultaneous motion in different directions).

The basic coordinate system of the machine is illustrated in Figure 12.13. The part program is identical for both front or rear turret machining. The difference is programmed in the "Setup" statement using the minor words "rear" or "front." Based on these instruments and following the turning program, the Compact II computer will create the list and tape files.

12.8.1 Example

Write a Compact II part program for the finish turning of a typical part as illustrated in Figure 12-14. The program will contain facing, contouring, grooving, threading, and drilling. As these operations are among the most common in part programs for turning centers, it is expected that they will provide a comprehensive overview of the Compact II programming process for turning.

12.8.2 Solution

Select the tool types and establish their dimensions as required for machining and programming. The Compact II tool statement requires us to specify the following dimensions:

TOOLn where *n* is the tool turret position

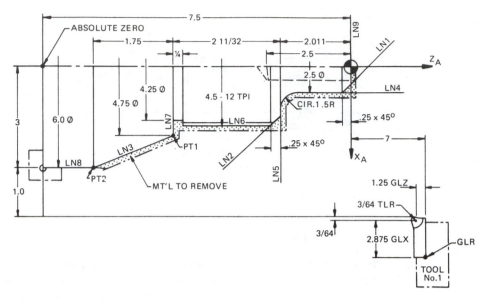

Figure 12.14 Typical turning part.

GLXn represents the gage length in the X-direction and *n* is the dimension from the Gage Length Reference Point (GLRP) to the center of the tool tip radius.

GLZn is the same requirement in the Z-direction

TLRn specifies the tool (nose) radius of value *n*.

The above requirements are written as follows:

ATCHG,TOOL1,OFFSET1,GLX2.875,GLZ1.25,TLR(3/64),0.008IPR,400FPM,RANGE1

The minor word RANGEn, used in conjunction with our first tool for facing, is only defined in the first tool statement. Its function is to generate the appropriate M code for a gear change and in our program *n* = 1. The balance of the minor words have already been explained in the milling example.

The tooling data can best be summarized in Table 12-1.

TABLE 12-1 TOOLING DATA

Operation	Tool No.	GLX	GLZ	TLR	Offset	IPR	FPM
Facing	1	2.875	1.25	3/64	1	0.008	400
Contouring	2	2.875	1.25	1/32	2	0.015	400
Grooving	3	2.5	1.0	0	3	0.004	80
Threading	4	2.75	1.0	0	4	—	50
Drilling	5	0	8.0	118TPA	5	0.006	80

Establish the part geometry by labeling points, lines, and circles on the part drawing, as shown in Figure 12.12.

Establish the program base and label the coordinate system on the part drawing. In programming for turning centers, the *XA* should always be at $X = 0$, while the *ZA* may be located at any length from the absolute zero. The latter should, however, be located on the part surface, as illustrated in Figure 12.12.

12.8.3 Writing the Part Program

12.8.3.1 Setup instruction

MACHIN,DEMOLATHE

where Demolathe is the name assigned to the link (postprocessor).

IDENT,CNC-JP-11 FRONT LATHE
SETUP,MOD2/1,CMOD2/2,ABSO2,ZEROS3,RPM2500,X3 + 1 + (3/64) + 2.875,
(continued)
Z7.5 + 7,LIMIT(X0/10,Z2/18),FRONT

The *X*- and *Z*- dimensions are established by the programmer. These dimensions should be far enough from the part to allow the operator to load and unload the parts.

ABSOn will cause the tape output to be in absolute mode, reflecting the tool point motions. $n = 2$ will generate a tape block containing a G50 code after each tool change.

ZEROSn will generate a machine control tape in decimal point format when $n = 3$ as programmed.

LIMIT specifies the absolute travel limits to be checked in relationship to the primary turret GLRP.

INIT, INCH/IN,INCH/OUT
BASE,XA,7.5ZA

12.8.3.2 Part geometry

DLN1,ZB,2.0D,45CCW
DLN2,4.5D,-2.011-0.25ZB,45CCW

The 4.5D,-2.011-0.25ZB defines a point. The computer will generate a line through this point, parallel to the Z-axis.

The 45CCW minor word will then rotate this same line 45° in the counterclockwise direction to form the required 45° chamfer.

DPT1,4.75D,-2.0-(11/32)-2.011ZB
DPT2,6D,-1.75-2.0-(11/32)-2.011ZB

DLN3,PT1,PT2
DLN4,2.5D
DLN5,2.011ZB
DLN6,4.5D
DLN7,-2.011-2.0-(11/32)ZB
DLN8,6.0D
DLN9,ZB
DCIR1,LN5/0.5ZL,LN4/0.5XL,0.5R

The reader has noticed, by now, the difference in the line definitions. The lines parallel to the Z-axis can be simply defined in terms of their diameter, right off the drawing.

12.8.3.3 Tool change and motion statements for facing

ATCHG,TOOL1,OFFSET1,GLX2.875,GLZ1.25,TLR(3/64),0.008IPR,400FPM,RANGE1
MOVEC,OFFLN1/0.35XL,OFFLN9/ZL

The MOVEC major word (Move to Cut) generates a rapid traverse motion with automatic deceleration to feed rate before the programmed point has been reached.

CUT,ONLN(XB),CSS/ON
CUT,0.05X,0.05Z

ONLN(XB) will move the tool point parallel to line 9, onto the center line of the part.

CSS/ON, Constant surface speed on, will generate the appropriate G code to increase the spindle rpm as the tool moves inward to the center of the part. This instruction is normally programmed when the machined contour changes in diameter.

12.8.3.4 Tool change and motion statements for contour turning

ATCHG,TOOL2,OFFSET2,GLX2.875,GLZ1.25,TLR(1/32),400FPM,0.015IPR
MOVEC,OFFLN1/XL,OFFLN9/0.1ZL
CUT,PARLN1,OFFLN4/XL,CSS/ON
ICON,CIR1,S(TANLN4),F(TANLN5),CCW
CUT,PARLN5,OFFLN2/XL
CUT,PARLN2,OFFLN6/XL
CUT,PARLN6,OFFLN7/ZL
CUT,PARLN7,OFFLN3/XL
CUT,PARLN3,OFFLN8/XL
MOVE,OFFLN9/0.1ZL

The above motion statements are easy to follow as they are directly related to the part geometry.

12.8.3.5 Tool change and motion statements for the turning of the groove

ATCHG,TOOL3,OFFSET3,GLX2.5,GLZ1.0TLR,CCS/OFF,80FPM,RPMD4.25,0.004IPR

This is yet another easy way of defining the spindle rpm in terms of the part diameter, 4.25, in conjunction with a surface speed of 80 fpm. The Compact II computer will substitute these values into the equation below:

$$n = \frac{80 \cdot 12}{4.25 \cdot \pi} = 71.9 \text{ rpm}$$

This spindle speed is reflected in the second tape block following the tool change in the tape file:

N0040 G97 S0071 M40

It is seen that the computer truncated 71.9 rpm to 71 in the S word.

```
MOVE,OFFLN3/0.05XL,OFFLN7/ZL
CUT,4.25D
CUT,OFFLN3/0.05XL,PARLN7
```

The "move" statement will rapid the tool point into cutting position. The Cut,4.25D will move the tool point, in feed, to 4.25 diameter, while the second cut statement will retract the tool point to clear the part.

12.8.3.6 Tool change and motion statements for thread cutting

```
ATCHG,TOOL4,OFFSET4,GLX2.75,GLZ1,TLR,50FPM,RPMD4.5
THRD12,S(LN5/0.3ZL),F(LN7/0.1ZL),MID(LN6/0.06XS),MAD(LN6),DEG29.5, (cont)
SDPTH0.012,FDPTH0.01,2SP/0.001,BDLP2,CYC4,OFF,CO
```

THRDn initiates the threading cycle, and $n = 12$ tpi.

MIDn represents the thread minor diameter and n the material to remove.

MADn is the thread major diameter, and n indicates its value.

DEGn specifies the angle n of infeed for the tool in the threading statement.

SDPTHn is the starting depth of the first cut, with $n = 0.012$.

FDPTHn is the final depth of the last cut and $n = 0.01$.

BDLPn requires a block delete to be generated for the last $n = 2$ passes.

CYC4/OFF will assure that the tool will not return to home position before the last pass has been completed.

nSP/i represents $n = 2$ spring passes at $i = 0.001$ between the roughing and the last passes.

In order to better appreciate the extent of the complexity of this last statement, the reader should carefully study the tape file following tool change 4.

12.8.3.7 Tool change and motion statements for drilling

```
ATCHG,TOOL5,OFFSET5,0.75TD,118TPA,GLX0,GLZ8,80FPM,0.006IPR
DRL,PT(XB,ZB),SDPTH0.75,FDPTH0.6,2.5DP,0.1CLEAR
```

SDPTHn specifies peck-drilling with $n = 0.75$ as first peck

FDPTHn specifies the subsequent pecking depth of $n = 0.6$.

```
END
```

It is strongly recommended that the reader study the source program in parallel with the list and the tape files to better understand the computer logic in processing the program.

The Source Program is the CNC computer program written by the programmer. This program consists of the start-up, geometry, tool change, and motion instructions. This program is punched on tape, read and stored on disk in an in-house Compact II computer. The program must then be processed and edited by the programmer off the terminal in an interactive conversational mode. The Compact II instruction word for processing is "Run," and the programmer answers questions from the computer by yes or no. In "run" mode, assisted by "substitute," the computer analyzes each statement for spelling, mathematical, and motion accuracy, etc. If the run is successful, the computer will print out the total machining time and the length of tape in both feet and meters.

```
MACHIN,DEMO   LATHE
IDENT, CNC-JP-11        FRONT LATHE
INIT,INCH/IN,INCH/OUT
SETUP,MOD2/1,CMOD2/2,ABSO2,ZEROS3,RPM2500,X3 + 1 + (3/64) + 2.875
Z7.5 + 7,LIMIT(X0/10,Z2/18),FRONT
BASE,XA,7.5ZA
DLN1,ZB,2D,45CCW
DLN2,4.5D, − 2.011 − .25ZB,45CCW
DPT1,4.75D, − 2 − (11/32) − 2.011ZB
DPT2,6D, − 1.75 − 2 − (11/32) − 2.011ZB
DLN3,PT1,PT2
DLN4,2.5D
DLN5, − 2.011ZB
DLN6,4.5D
DLN7, − 2.011 − 2 − (11/32)ZB
DLN8,6D
DLN9,ZB
DCIR1,LN5/.5ZL,LN4/.5XL,.5R
ATCHG,TOOL1,OFFSET1,GLX2.875,GLZ1.25,TLR(3/64),.008IPR,400FPM,RANGE1
MOVEC,OFFLN1/.35XL,OFFLN9/ZL
CUT,ONLN(XB),CSS/ON
CUT,.05X,.05Z
```

```
ATCHG,TOOL2,OFFSET2,GLX2.875,GLZ1.25,TLR(1/32),400FPM,.015IPR
MOVEC,OFFLN1/XL,OFFLN9/.1ZL
CUT,PARLN1,OFFLN4/XL,CSS/ON
ICON,CIR1,S(TANLN4),F(TANLN5),CCW
CUT,PARLN5,OFFLN2/XL
CUT,PARLN2,OFFLN6/XL
CUT,PARLN6,OFFLN7/ZL
CUT,PARLN7,OFFLN3/XL
CUT,PARLN3,OFFLN8/XL
MOVE,OFFLN9/.1ZL
ATCHG,TOOL3,OFFSET3,GLX2.5,GLZ1,TLR,CSS/OFF,80FPM,RPMD4.25,.004IPR
MOVE,OFFLN3/.05XL,OFFLN7/ZL
CUT,4.25D
CUT,OFFLN3/.05XL,PARLN7
ATCHG,TOOL4,OFFSET4,GLX2.75,GLZ1,TLR,50FPM,RPMD4.5
THRD12,S(LN5/.3ZL),F(LN7/.1ZL),MID(LN6/.06XS),MAD(LN6),DEG29.5,SDPTH.012
FDPTH.01,2SP/.001,BDLP2,CYC4/OFF,CO ,
ATCHG,TOOL5,OFFSET5,.75TD,118TPA,GLZB,GLX0,80FPM,.006IPR
DRL,PT(XB,ZB),SDPTH.75,FDPTH.6,2.SDP,.1CLEAR
END
```

The Tape File is produced at the same time, provided there are no errors during processing. This file should be printed and checked prior to punching the CNC tape. Changes to this file can be made in "edit" mode, thus providing the knowledgeable programmer with an additional level of flexibility. Once the programmer is satisfied with the output, the computer can be instructed to punch out the machine control tape. The programmer has the option to punch the tape, punch and print the file or print only.

```
                    TAPE
                    TAPE FILE: /CNC-JP11T/ OK? YES
                    EIA? NO
                    OUTPUT TO: TPT
                    TURN PUNCH ON,HIT CR

                    N0001G90
                    N0002G94
                    N0003G04X2.T0101
                    N0004G50XS.0938Z13.25
                    N0005G97S0520M40
                    N0006G01X3.0288Z7.5469F300.M03
                    N0007G50S0625
                    N0008G96R1.4676S0400
                    N0009G95
                    N0010G01X.F.008
                    N0011X.1Z7.5969
                    N0012G97S0625
```

```
N0013G94
N0014G01X8.0938Z13.25F300.
N0015G90
N0016G94
N0017G04X2.T0202
N0018G50X8.0938Z13.25
N0019G97S0625M40
N0020G01X1.826Z7.6313F300.M03
N0021G50S0625
N0022G96R.8818S0400
N0023G95
N0024G01X2.5626Z7.2631F.015
N0025Z5.989
N0026G02X3.5Z5.5203I.4687K
N0027G01X4.0258
N0028X4.5626Z5.252
N0029Z3.1765
N0030X4.794
N0031X6.0626Z1.4006
N0032G00Z7.6313
N0033G97S0254
N0034G94
N0035G01X8.0938Z13.25F300.
N0036G90
N0037G94
N0038G04X2.T0303
N0039G50X8.8438Z13.5
N0040G97S0071M40
N0041G01X4.8562Z3.1453F300.M03
N0042G95
N0043G01X4.25F.004
N0044X4.8562
N0045G94
N0046G01X8.8438Z13.5F300.
N0047G90
N0048G94
N0049G04X2.T0404
N0050G50X8.3438Z13.5
N0051G97S0062M40
N0052G01X4.6Z5.789F300.M03
N0053X4.4814Z5.7554F.25
N0054G33Z3.2453K.08333
N0055X4.6I.08333
N0056G00Z5.789
N0057G01X4.4624Z5.7501
N0058G33Z3.2453K.08333
N0059X4.6I.08333
N0060G00Z5.789
```

```
N0061G01X4.4432Z5.7446
N0062G33Z3.2453K.08333
N0063X4.6I.08333
N0064G00Z5.789
N0065G01X4.4238Z5.7391
N0066G33Z3.2453K.08333
N0067X4.6I.08333
N0068G00Z5.789
N0069G01X4.404Z5.7336
N0070G33Z3.2453K.08333
N0071X4.6I.08333
N0072G00Z5.789
N0073G01X4.384Z5.7279
N0074G33Z3.2453K.08333
N0075X4.6I.08333
N0076G00Z5.789
N0077G01X4.382Z5.7273
N0078G33Z3.2453K.08333
N0079X4.6I.08333
N0080G00Z5.789
N0081G01X4.38Z5.7268
N0082G33Z3.2453K.08333
N0083X4.6I.08333
N0084G90
N0085G94
N0086G01X8.3438Z13.5F300.
N0087G90
N0088G94
N0089G04X2.T0505
N0090G50X13.8438Z6.5
N0091G97S0407M40
N0092G00X13.8438Z7.6M03
N0093X.
N0094G95
N0095G01Z7.0098F.006
N0096G00Z7.6
N0097Z7.0598
N0098G01Z6.4922
N0099G00Z7.6
N0100Z6.5422
N0101G01Z5.9472
N0102G00Z7.6
N0103Z5.9972
N0104G01Z5.3747
N0105G00Z7.6
N0106Z5.4247
N0107G01Z4.7747
N0108G00Z7.6
```

```
N0109G94
N0110G01X13.8438Z6.SF300.
N0111G04X1.T0500
N0112M30
```

12.9 APT PART PROGRAMMING

The APT programming example will show the minor differences between APT and Compact II. The source statements should appear familiar, and new formats will be clarified as part of the program Comments.

Task.   Write an APT part program for milling the multi-pocket component illustrated in Figure 11.1, p. 183.

Assumptions.

Tool No. 1: 0.500″ dia × 4″ Center drill
Tool No. 2: 0.375″ dia × 4″ Twist drill
Tool No. 3: 0.500″ dia × 4″ End mill

Solution.

1	PARTNO SAMPLE-1 G04 10 ISS. NC JS	
2	$$ ACCNT	Programmer's account number.
3	SYN/AA, ATANGL, CA, CANON, CT, CENTER. C, CIRCLE, . . . (etc)	Instruction allows the programmer to use the abbreviated (short) version instead of the full-length APT word.
4	SYN/ZL, ZLARGE, ZS, ZSMALL, FR, FEDRAT, IP, INDIRP	As above, ZL to replace ZLARGE, etc.
5	MACHIN/MACAX4, 2	Postprocessor No., OKK 500.
6	M1 = MX/TRANSL, −2, −2, 0	Defining a Matrix, MX, for the part's coordinate system. The matrix specifies the relation between two coordinate systems as MATRIX/P1,P2,P3 for translation only. Operator must set Z_s to top of part. APT has approximately 13 Matrix definition versions.
7	SETPT = P/0, 0, 0	Position of set point.
8	CLPRNT/OPTION	Instructs system to print out the cutter center locations.
9	PPRINT *** PART GEOMETRY ***	PPRINT statements are comments.
10	TOOLNO/01, DRILL, .5, 4.0	Defining the center drill.
11	TOOLNO/02, DRILL, 0.375, 4.0	Defining the twist drill.
12	PPRINT *** PART GEOMETRY ***	Comment for the printout.
13	TH1 = P/−1.2, −0.75, 0.1	Location of Tooling Hole 1.

14	TH2 = P/−17.0, −3.5, 0.1	Location of Tooling Hole 2.
15	P0 = P/−3, −2.375	Location of pocket coordinate system.
16	P1 = P/−(3−0.3125),− (1.625 +0.75+0.75), 0.1	Definition of point 1, including necessary calculations.
17	P2 = P/−3, −1.625	Point 2.
18	C1 = C/CT, P1,R, 0.3125	Circle 1.
19	C2 = C/CT, P2, R, 0.625	Circle 2.
20	L0 = L/XAXIS	Line 0 is the X-axis.
21	L1 = L/P0, RIGHT, T, C1	Line 1, starting at P0, tangent to the right side of circle C1.
22	L2 = L/RIGHT, T, C1, RIGHT, T, C2	Line 2, tangent on right to circles C1 and C2.
23	L3 = L/(P/−3.625,−3), LEFT, T, C2	Line 3, starting at point defined in brackets, tangent to left side of circle C2.
24	L4 = L/PA, L0, YS, 2.375	Line 4.
25	C3 = C/XL, L3, YL, L4, R, 0.5	Circle 3.
26	PL0 = PL/0, 0, 1, 0	Plane 0.
27	PL1 = PL/0, 0, 1, −0.25	Plane 1. One of the many optional formats to define planes, PLANE/ a,b,c,d, where a,b, and c are direction cosines, and d is the perpendicular distance from the plane to origin of the coordinate system.
28	PR/0	Skip to the next page in print output only.
29	PRINT ***_____***	
30	PRINT *** OPERATOR TO BOLT BLANK PER SETUP***	
31	PRINT ***SHEET OR ANY OTHER SHOP INSTRUCTION ***	
32	PRINT ******************	
33	$$ ** MOTION STATEMENTS TO FOLLOW **	Comment for programmer use.
34	FROM/SETPT	Normally used for initial position of cutter.
35	LOADTL/01, AT, 5, 5	Tool change for Tool 1 at location specified.
36	LOCKX	Lock table.
37	TR/M1	Transforms cutter location macro 1.
38	CUTTER/0.5	
39	SPINDL/1400	
40	RAPID, GT/(P/TH1, CA, , ,4)	Rapid to TH1 location. No change on X or Y; Z = 4.
41	JOE = MACRO/Q1, Q2	Start of definition for macro "JOE." Q1 and Q2 are variables.
42	RAPID, GT/Q1	Rapid Go To Q1 = TH1.
43	GD/−Q2,6	Drill to −Q2 = 0.25 @ 6 IPM.
44	RAPID, GD/Q2	Rapid Go Delta to Q2.
45	TERMAC	Conclusion of macro definition.
46	CALL/JOE, Q1 = TH1, Q2 = 0.25	This statement activates macro "JOE" for the drilling of the first tooling hole.
47	CALL/JOE,Q1 = TH2, Q2 = 0.25	Drill 2nd tooling hole.
48	RAPID, GO/3	

49	$$ PR/0	
50	PPRINT *** DRILL TWO TOOLING HOLES ***	This statement will print a comment in the tape file.
51	LOADTL/02, AT, 5,5	
52	CUTTER/0.375	Drill diameter = 0.375″.
53	SPINDL/1800	Start spindle, 1800 rpm.
54	RAPID,GT/(P/TH1, CA, , ,4)	Rapid to TH1.
55	COOLNT/FLOOD	
56	CALL/JOE, Q1 = TH1, Q2 = 0.8	Drill through.
57	CALL/JOE,Q1 = TH2, Q2 = 0.8	Drill through.
58	COOLNT/OFF	Turn coolant off.
59	RAPID, GD/3	
60	PR/0	
61	LOADTL/03, AT, 5, 5	
62	STOP	
63	PPRINT * INSTALL 2 − 5/16 SHOULDER BOLTS *	
64	PPRINT * REMOVE CLAMPS *	
65	CUTTER/0.5,0	Cutter diameter and bottom radius.
66	SPINDL 1800	
67	RAPID, GT/(P/P1, CA, , − 3.4)	Note: Between ", ," as there is no data, the system inserts 0.
68	INDEX/1	Start of pocket macro.
69	RAPID, GT/P1	
70	TH/0, 0.03	This statement specifies additional offset to surfaces in the format THICK/PART SURF, OFFSET. Do not misinterpret TH for a Tooling Hole.
71	GO/ON, (L/PA, L4, YL 0.75), PL1, ON, (L/P1, PA, L1), 10	Cut down to pocket bottom in the format LINE/PARLEL, LINE4, YLARGE, BY 0.75″, PLANE 1, ON LINE/POINT1, PARLEL, LINE 1, 10 IPM.
72	TLON, GL/(L/PA, L4, YL, 0.75), 25	
73	GL.L1	
74	TLLFT, GL/L4	
75	GR/L1	ROUGH
76	GF/C1	
77	GF/L2	CUT
78	GF/C2	
79	GF/L3	POCKET
80	GF/C3	
81	GF/L4, PAST, L1	
82	GD/0.01	
83	TH/0.002, 0	
84	GO/L1, PL1, L4, 20	
85	TLLFT, GL/L1	FINISH
86	GF/C1	
87	GF/L2	MILL
88	GF/C2	
89	GF/L3	POCKET

90	GF/C3	
91	GF/L4,PAST,L1	
92	GD/0.35, 50	
93	INDEX/1, NM	NOTE: The above motion statements, from INDEX/1, line 68, to the end of the macro, INDEX/1, NM (No More), will be used for machining the other 3 pockets in a copy mode, in the format of the next statement.
94	COPY/1,TRANSL, −4.5, 0, 0, 1	Machine Pocket No. 2. Translation, X−4.5, Y0, Z0, 1 time.
95	COPY/1,TRANSL, −8.5, 0, 0, 1	Machine Pocket No. 3. Translation, X−8.5, Y0, Z0, 1 time.
96	COPY/1,TRANSL, −12.75, 0, 0, 1	Machine Pocket No. 4. Translation, X−12.75, Y0, Z0, 1 time.
97	COOLNT/OFF	
98	TH/0	Reset thickness deviation to zero.
99	RAPID, GT/−15.5, −2.125, 4	
100	TR/NM	Turn transformation off.
101	END	
102	PR/3, ALL	This statement will cause the printing out at execution of the canonical form of each surface and scalar variable.

Conclusions.   It is still the responsibility of the programmer to know and specify the cutting and process parameters, hence to be familiar with processes, tools, and computer hardware. The programmer should also select, as applicable, mainframes or PCs (APT is now available on microcomputers).

12.9.1 Complete Listing of APT Source Program

```
 1. PARTNO SAMPLE-1-604-10     ISS. NC.
 2. $$            ACCNT. - AB -
 3. SYN/AA,ATANGL,CA,CANON,CT,CENTER,C,CIRCLE,CY,CYLNDR,D,DELAY
 4. SYN/GB,GOBACK,GD,GODLTA,GF,GOFWD,GL,GOLFT,GR,GORGT,GT,GOTO
 5. SYN/IV,INDIRV,I,INTOF,LG,LARGE,LC,LCONIC,L,LINE,MX,MATRIX,NM,NOMORE
 6. SYN/O,OBTAIN,PA,PARLEL,PE,PERPTO,PL,PLANE,P,POINT,PR,PRINT,R,RADIUS
 7. SYN/SM,SMALL,S,STOP,T,TANTO,TH,THICK,TA,TLAXIS,N,TMARK,TR,TRACUT
 8. SYN/V,VECTOR,XL,XLARGE,XS,XSMALL,YL,YLARGE,YS,YSMALL
 9. SYN/ZL,ZLARGE,ZS,ZSMALL,FR,FEDRAT,IP,INDIRP
10. MACHIN/MACAX4,2
11. M1      = MX/TRANSL, −2, −2,0        $$ SET  -Z-  TO TOP OF PART
12. SETPT   = P/0,0,0
13. CLPRNT/OPTION
14. PPPRNT
15. TOOLNO/01,DRILL,.500,4.00                     $$ .500 DIA CTR DRILL
16. TOOLNO/02,DRILL,.375,4.00                     $$ .375 DIA DRILL
17. TOOLNO/03,MILL,.500,4.00                       $$ .500 DIA, O CR. E.M.
18. PPRINT     ****            PART GEOMETRY        ****
19. TH1      = P/−1.2, −.75,.1
20. TH2      = P/−17, −3.5,.1
21. P0       = P/−3, −2.375
```

```
22. P1       = P/−(3−.3125),−(1.625+.75+.75),.1
23. P2       = P/−3,−1.625
24. C1       = C/CT,P1,R,.3125
25. C2       = C/CT,P2,R,.625
26. L0       = L/XAXIS
27. L1       = L/P0,RIGHT,T,C1
28. L2       = L/RIGHT,T,C1,RIGHT,T,C2
29. L3       = L/(P/−3.625,−3),LEFT,T,C2
30. L4       = L/PA,L0,YS,2.375
31. C3       = C/XL,L3,YL,L4,R,.5
32. PL0      = PL/0,0,1,0
33. PL1      = PL/0,0,1,−.25
34. PR/0
35. PPRINT   ********************************************************************
36. PPRINT   ***               MOUNT BILLET AS PER SET-UP SHEET              ***
37. PPRINT   ****                        CLAMP                              ****
38. PPRINT    ***             CENTRE DRILL TOOLING HOLES                    ***
39. PPRINT   ********************************************************************
40. FROM/SETPT
41. LOADTL/01,AT,5,5
42. LOCKX
43. TR/M1
44. CUTTER/.500
45. SPINDL/1400
46. RAPID,GT/(P/TH1,CA,,,4)
47. JOE = MACRO/Q1,Q2
48. RAPID,GT/Q1
49. GD/−Q2,6
50. RAPID,GD/Q2
51. TERMAC
52. CALL/JOE,Q1=TH1,Q2=.25
53. CALL/JOE,Q1=TH2,Q2=.25
54. RAPID,GD/3
55. $$ PR/0
56. PPRINT        ******        DRILL TWO TOOLING HOLES        ******
57. LOADTL/02,AT,5,5
58. CUTTER/.375
59. SPINDL/1800
60. RAPID,GT/(P/TH1,CA,,,4)
61. COOLNT/FLOOD
62. CALL/JOE,Q1=TH1,Q2=.8
63. CALL/JOE,Q1=TH2,Q2=.8
64. COOLNT/OFF
65. RAPID,GD/3
66. PR/0
67. LOADTL/03,AT,5,5
68. STOP
69. PPRINT   ********************************************************************
```

70. PPRINT   ***   INSTALL TWO .375 SHOULDER BOLTS, REMOVE CLAMPS   ***
71. PPRINT   ****   MILL POCKETS   ****

72. CUTTER/.500,0
73. SPINDL/1800
74. RAPID,GT/(P/P1,CA,,−3.4)
75. INDEX/1
76. RAPID,GT/P1
77. TH/0,.03
78. GO/ON,(L/PA,L4,YL,.75),PL1,ON,(L/P1,PA,L1),10
79. TLON,GL/(L/PA,L4,YL,.75),25
80.   GL/L1
81. TLLFT,GL/L4
82.   GR/L1
83.   GF/C1
84.   GF/L2
85.   GF/C2
86.   GF/L3
87.   GF/C3
88.   GF/L4,PAST,L1
89. GD/.01
90. TH/.002,0
91. GO/L1.PL1,L4,20
92. TLLFT,GL/L1
93.   GF/C1
94.   GF/L2
95.   GF/C2
96.   GF/L3
97.   GF/C3
98.   GF/L4,PAST,L1
99. GD/.35,50
100. INDEX/1,NM
101. COPY/1,TRANSL,−4.5,0,0,1
102. COPY/1,TRANSL,−8.5,0,0,1
103. COPY/1,TRANSL,−12.75,0,0,1
104. COOLNT/OFF
105. TH/0
106. RAPID,GT/−15.5,−2.125,4
107. TR/NM
108. END
109. PR/3,ALL

12.9.2 Machine Control Unit Data (Tape File)
Corresponding to APT Source Program

```
N001 G00 G46 G91 X0 H47
N002 G46 Y0 H48
N003 G92 X0 Y0 Z0
```

```
N004 G90
N005 X50000 Y50000
N006 T01
N007 M06
N008 G92 X50000 Y50000 Z0 A0
N009 G90
N010 M78
N011 M03 S1400
N012 G00 G43 H01 X-32000 Y-27500 Z40000
N013 G00 Z10000
N013 G01 Z-1500 F60
N015 G00 Z1000
N016 G00 X-190000 Y-55000
N017 G01 Z-1500
N018 G00 Z1000
N019 G00 Z31000
N020 G00 G60 Z0
N021 X50000 Y50000
N022 M06
N023 T02
N024 M06
N025 G92 X50000 Y50000 Z0 A0
N026 G90
N027 M03 S1800
N028 G00 G43 H02 X-32000 Y-27500 Z40000
N029 M08
N030 G00 Z1000
N031 G01 Z-7000
N032 G00 Z1000
N033 G00 X-190000 Y-55000
N034 G01 Z-7000
N035 G00 Z1000
N036 M09
N037 G00 Z31000
N038 G00 G60 Z0
N039 X50000 Y50000
N040 M06
N041 T03
N042 M06
N043 G92 X50000 Y50000 Z0 A0
N044 G90
N045 M00
N046 M03 S1800
N047 G00 G43 H03 X-46875 Y-50000 Z40000
N048 G00 Y-51250 Z1000
N049 G01 X-47175 Y-36550 Z-2500 F100
N050 X-49700 F250
```

```
                    N051 Y-40950
                    N052 X-47200
                    N053 Y-51250
                    N054 G17
                    N055 G03 X-46550 I325 F214
                    N056 G01 Y-36250 Z-2500 F250
                    N057 G03 X-53450 I-3450 F250
                    N058 G01 Y-38750 Z-2500 F250
                    N059 G03 X-51250 Y-40949 I2200 F250
                    N060 G01 X-47200 Y-40950 Z-2500 F250
                    N061 Z-2400
                    N062 X-47500 Y-41250 Z-2480 F200
                    N063 Y-51250
                    N064 G03 X-46250 I625 F200
                    N065 G01 Y-36250 Z-2480 F200
                    N066 G03 X-53750 I-3750 F200
                    N067 G60 Y-38750 Z-2480 F200
                    N068 G03 X-51250 Y-41250 I2500 F200
                    N069 G01 X-47500 Z-2480 F200
                    N070 Z1020 F500
                    N071 G00 X-91875 Y-51250 Z1000
                    N072 G601 X-92175 Y-36550 Z-2500 F100
                    N073 X-94700 F250
                    N074 Y-40950
                    N075 X-92200
                    N076 Y-51250
                    N077 G03 X-91550 I325 F214
                    N078 G01 Y-36250 Z-2500 F250
                    N079 G03 X-98450 I-3450 F250
                    N080 G01 Y-38750 Z-2500 F250
                    N081 G03 X-96250 Y-40949 I2200 F250
                    N082 G01 X-92200 Y-40950 Z-2500 F250
                    N083 Z-2400
                    N084 X-92500 Y-41250 Z-2480 F200
                    N085 Y-51250
                    N086 G03 X-91250 I625 F200
                    N087 G01 Y-36250 Z-2480 F200
                    N088 G03 X-98750 I-3750 F200
                    N089 G01 Y-38750 Z-2480 F200
                    N090 G03 X-96250 Y-41250 I2500 F200
                    N091 G01 X-92500 Z-2480 F200
                    N092 Z1020 F500
                    N093 G00 X-131875 Y-51250 Z1000
                    N094 G01 X132175 Y-36550 Z-2500 F100
                    N095 X-134700 F250
                    N096 Y-40950
                    N097 X-132200
```

```
N098 Y-51250
N099 G03 X-131550 I325 F214
N100 G01 Y-36250 Z-2500 F250
N101 G03 X-138450 I-3450 F250
N102 G01 Y-38750 Z-2500 F250
N103 G03 X-136250 Y-40949 I2200 F250
N104 G01 X-132200 Y-40950 Z-2500 F250
N105 Z-2400
N106 X-132500 Y-41250 Z-2480 F200
N107 Y-51250
N108 G03 X-131250 I625 F200
N109 G01 Y-138750 Z-2480 F200
N110 G03 X-138750 I-3750 F200
N111 G01 Y-38750 Z-2480 F200
N112 G03 X-136250 Y-41250 I2500 F200
N113 G01 X-132500 Z-2480 F200
N114 Z1020 F500
N115 G00 X-174375 Y-51250 Z1000
N116 G01 X-174675 Y-36550 Z-2500 F100
N117 X-177200 F250
N118 Y-40950
N119 X-174700
N120 Y-51250
N121 G03 X-174050 I325 F214
N122 G01 Y-36250 Z-2500 F250
N123 G03 X-180949 I-3450 F250
N124 G01 X-180950 Y-38750 Z-2500 F250
N125 G03 X-178750 Y-40949 I2200 F250
N126 G01 X-174700 Y-40950 Z-2500 F250
N127 Z-2400
N128 X-175000 Y-41250 Z-2480 F200
N129 Y-51250
N130 G03 X-173750 I625 F200
N131 G01 Y-36250 Z-2480 F200
N132 G03 X-181250 I-3750 F200
N133 G01 Y-38750 Z-2480 F200
N134 G03 X-178750 Y-41250 I2500 F200
N135 G01 X-175000 Z-2480 F200
N136 Z1020 F500
N137 G00 Z40000
N138 G00 G60 Z0
N139 M06
N140 G60 X0 Y0
N141 M05
N142 M09
N143 M02
```

12.10 CAD-CAM PART PROGRAMMING

There are many interactive part programming systems for computer numerically controlled machine tools. Some of them are based on microcomputers, and their price-performance ratio is remarkable. The others are based mainly on large minicomputers, and while they all require study and practice, they represent a quantum leap from conventional computer-assisted part programming. These CAM-driven programming methods can provide an in-process graphic display of the geometries and tool motions. As a result, the language programming requirement is eliminated.

In the following pages, an overview of one of these systems, MDSI Schlumberger's BRAVO3, specifically the BRAVO3 Geometric Modeling, is described. The BRAVO3 system has full five-axis programming capabilities, and the programmer can make use of both an APT-like text, and/or graphic input mode. The user is guided through the programming process by selecting command items from a "marching" menu, and the prompts can be answered by use of keyboard, tablet, or cursor controller.

BRAVO3 is called up on the Digital Equipment VAX-VMS computer, using the command word GEM, short for **GE**ometric **M**odeler. BRAVO3 is a three-dimensional (3D) interactive part modeling system, using menus and prompts. GEM uses a text flashing cursor and a graphics crosshair cursor when expecting an input from the programmer.

Directions involve YES/NO questions, SELECT for menu times, PICK for graphics display geometric elements, and ENTER for information typed at the keyboard. Instructions SELECT-ed or PICK-ed are implemented using the cursor controller, i.e., pressing the upright pen straight down on the tablet for an instant. Instructions ENTER-ed are completed by pressing the carriage return.

Almost totally menu-based, BRAVO3 has an INITIAL menu, located on the bottom left side of the screen, used with a system prompt. A FIXED menu is located across the bottom of the screen, and a COLOR bar is provided at the far right. File storage, retrieval, and manipulation is based on the VAX/VMS operating system, and can be accessed at VMS level, or within the BRAVO3 system.

Unlike APT or Compact II, BRAVO3 uses a graphics approach to define the part geometry. Its built-in efficiency is increased even more when the geometric elements are assigned to the same work plane that the tool path will use. The part model should be developed in sequence, from element to connecting element, rather than defining, say, all the points, then all the lines, all the circles, etc., as is customary in conventional computer-assisted programming. Unneeded element segments can be trimmed, and construction is facilitated by extensive use of ZOOM. The screen image is positioned using CENTER, and all the necessary construction lines may be located in appropriate construction LAYERS for as long as needed. HELP is available in both screen and VAX modes.

Two-Dimensional (Planar) Geometry.   POINTS are defined by selecting TRAP, through absolute XYZ coordinates, or in relation to existing predefined geometry:

- As the INTERSECTION of two intersecting geometric elements,
- AROUND an arc or circle at a specified angle,
- RELATIVE to an existing point,
- INCREMENTAL, i.e., RELATIVE to the last point defined,
- At ORIGIN (0,0,0).

TRAP locations are on the point for a POINT, at center and endpoints for a LINE, at center for a CIRCLE, and endpoints and radial center for an ARC. LINES, CIRCLES, and ARCS (including FILLETS) are defined utilizing the same geometric principles used by computer-assisted programming, with the major difference that graphic methods are applied on the screen. SPLINES will generate a mathematically simulated curve through graphic selection of a number of specified points. WORK PLANES are used to define geometry groups in a local XYZ coordinate system with a common ORIGIN point. All the mathematical work, and related "number crunching" is transparent to the user. The data calculated is stored and can be accessed if needed at any time.

Three-Dimensional Geometry.   Points, lines, and arcs are defined in a manner very similar to the one used in planar geometry. The circle definition is extended to the intersection of a sphere with a plane.

SURFACES are based on "primitives," such as spheres, cylinders, and cones, each defined graphically using on-screen menus. The purpose of vectors (VEC) is to help with directions, and they are specified off the GET VECTOR menu. Added definition flexibility is given by the SWEPT menu. The RULED menu provides surface definition by straight-line connection of cross sections. The WARPED menu defines surfaces as spline connection of cross sections.

Added graphics flexibility is given by MOVE/COPY, through TRANSLATION, ROTATION, MIRROR image, SCALING, etc. Efficiency is enhanced by LAYERS, which can be created, deleted, activated, renamed, and turned on or off as required.

FILES can be handled from within the graphics menu. Files can be LOADED, SAVED, APPENDED, CLEARED, while DRAFT/SHAPE, SHAPE VIEW, and SURFACE facilitate the use of the graphics definitions with the *BRAVO3 complex surface machining system.* MACRO allows efficient file manipulation, by use of BREAK, REFERENCE, or RE-REFERENCE.

Screen images or menus can be controlled by using a FIXED MENU, located across the bottom of the screen. This includes MENU, SCREEN, and PROCESSING functions, with instructions such as RETURN, HELP, ZOOM, CENTER, RVIEW (R = restore), SVIEW (S = save), VIEW (ROTATE VIEW, VIEW ALL, REPAINT, RESIZE VIEW, etc.); VISIBLE, MODIFY, MODALS (group color, axes display, units, etc.); GEOMETRIC DATA (which allows for retrieval of all calculated geometric information), etc.

Lastly, the functions of the VAX/VMS operating system allow the programmer, working at the $ prompt, to handle files (COPY, DELETE), examine all or specific parts of the DIRECTORY, set appropriate PROTECTION, PURGE duplicate files, and perform any other required functions.

Operation.   The programmer, using a graphics terminal, can create or transport an existing engineering part drawing into the programming session. Schlumberger has made tremendous strides toward making the software user-friendly and easy to use for anyone familiar with CNC programming and terminology. See Figure 12.15. For the obvious command choices, the system prompts the user for [Yes/No] Yes? answers. For example, Front (Rear) Turret ? If the answer is Yes, i.e., Front, all that is required is keying Return (↵). If the answer is No, type the word No and key in Return (↵). Please note that Return is sometimes shown as CR (Carriage Return).

SELECT

When the system prompt is Select, the programmer must choose a menu command, using a tablet or a mouse, such as

TOOL SIZE

RADIUS
DIAMETER
WIDTH

Selection takes place by positioning the cursor over the desired menu option, then pressing down on the light pen. The appropriate menu item will be highlighted by a bright border.

PICK

The prompt requires the programmer to identify an existing element, normally graphic (point, line, toolpath, circle, boundary, etc.), but possibly text. When the prompt has been satisfied, the system responds by flashing the element picked. Incorrect picks can easily be unpicked. The desired areas can be enlarged (zoomed) prior to picking.

ENTER

The prompt requires the typing of an alphanumeric command from the keyboard,

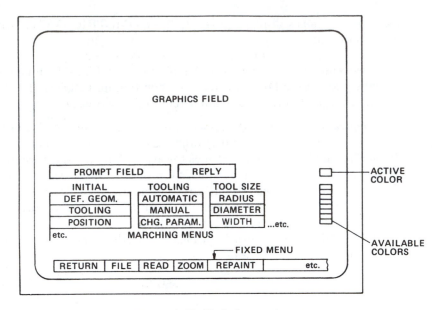

Figure 12.15 Typical screen layout.

then keying the CR (↵). Mistakes can be easily corrected using Backspace/Delete or Rubout keys. If exact values are unavailable, arithmetic expressions may be used, such as LENGTH : 3.71 ∗ SIN[30].

The user-friendliness is enhanced by an easy screen control that allows the programmer to move the center of the working screen and zoom as required. Views can be displayed in XY, YZ, ZX, isometrics, etc. The user can eliminate costly errors in toolpath by tilting or rotating the graphic display.

12.10.1 BRAVO3 Input/Output

It is assumed that the part programmer has an existing design or engineering drawing.

Initial Input.   This is relevant to a specific CNC system, such as

MACHIN, 9400, 57090

where the first number refers to the machine type, in this case an SL-3, and the second is the link number.

IDENT, SAMPLE
INIT, INCH/IN, INCH/OUT
SETUP, 5X . . . . etc.

(See sample program at end of chapter.) The system uses an on-line Machine Tool Driver (MTD).

Main Program.   The information is supplied interactively by the CNC programmer. It addresses part geometry, tooling, motion, and other auxiliary details. This section will generate a SOURCE PROGRAM, previously written by the part programmer. An extract from the sample program is shown below:

```
DLN1,SLAYER,PT(ZB-4,XBO),PT(ZBO,XBO)
. . . . .
DPT14,PLAYER1,ZBO,XBO
. . . . .
DCIR5,SLAYER2,R.2,TANELM(PT. . . .etc.
```

(See sample program.)

This source is stored in a file for later use. It can be run at any future time to display the part or any engineering changes.

Tape File.   The tape file is the actual result of a processed List File through a machine tool link (MTD), in the case of this program, 57090. This file is stored separately, ready for use in cutting, machining, or fabricating, as the case may be. It can be transferred to the CNC machine in the shop through punched tape or direct downloading through a network.

The above overview should help in working through the sample program below.

12.10.1.1 Sample program

Task.   Write a BRAVO3 part program for the turning of the part illustrated in Figure 12.16.

Assumptions.

- The drawing is not dimensioned in order not to clutter the process.
- The part is located in a database called "sample."
- The part filename is "turn."
- The tools involved are called
     Tool No. 1
     Tool No. 2
     Tool No. 3
- The graphics shown in Figure 12.17 are displayed on the work station screen.

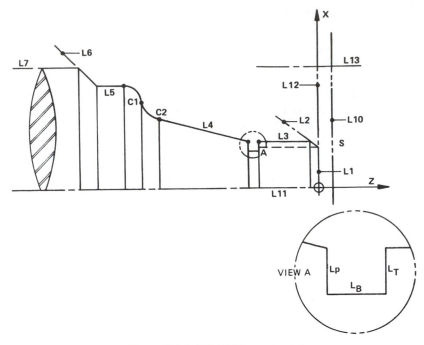

Figure 12.16 BRAVO3 sample part.

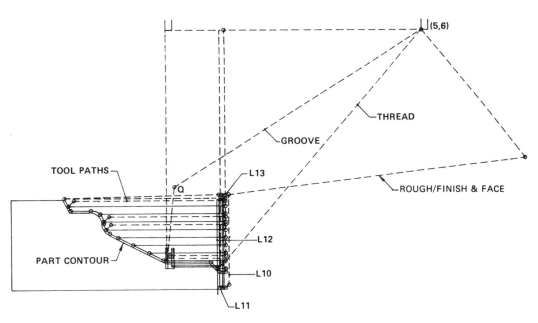

Figure 12.17 BRAVO3 tool path.

12.10.2. BRAVO3 SOLUTIONS

(a.) Display the part. Select the following commands from the menus:

APPLICATION MECHANICAL	DESIGN GEOM. MODELER
EDITOR -DB	From existing database.
OLD-PROMPT	Will display all the names of drawings.
SAMPLE	From the SAMPLE database.
TURN	Display the graphics of the part called "turn".

If the name of the database and part are known, the programmer can pick OLD-NAME instead of OLD-PROMPT, and type in the names SAMPLE and TURN on the keyboard. At this point, the geometric modeler can be left using the command

EXIT

(b.) Enter NC software and provide required tooling data.

MECHANICAL	From the initial menu.
MANUFACTURING	The manufacturing options are displayed.
GENERAL-NC	The NC options are displayed.

The first option represents the requirements for setup.

NC-SETUP	This will allow the programmer to answer the prompts for turning the sample part.
OLD-PROMPT	This will display the list of CNC machines in the shop.
LATH	This is a name of a home-written subprogram that will insert the following start-up statements into the lathe source program:
	MACHIN, 9400, 57090
	IDENT
	INIT, INCH/IN, INCH/OUT
	SETUP,5X,4Z,CMOD(6,1,1),MOD(3,1),ZEROS3.

These statements are suitable for most programs, and any deviations may be edited as required. This subprogram eliminates a number of prompts, with the possibility of either improper data or improper order of the data.

GEOMETRIC MODELER	
OLD-PROMPT-TURN	
DONE-DONE	This completes the setup part of the interactive program.
(prompt)	(reply)
SAVE PART PROGRAM ON:	TURN1⏎

The source file will be saved as TURN1

(prompt)   (reply)
ENTER USER DATA: DEMO TURNING JULY 22 ↵
                                  This note will be punched in the leader portion of the tape using
                                  man-readable characters. See third line of source program.

Create the part boundary for rough turning. NOTE: If the axes are displayed, they should be turned off as:

OTHER-MODALS-AXES   Commands will start with the fixed menu
REPAINT   Refresh graphics.

The color should be changed, so that the part boundary can be distinguished from the rest of the part geometry. After a new color was picked from the color box, the program can proceed.

GEOMETRY-BOUNDARY-ELEMENT

                                  Pick L1, L2, L3, L4, C2, C1, L5, L6, L7.
TERMINATE   End of picking.

The system will now highlight the part boundary using the new color. It should be noted that the first and last elements, L1 and L7, are not constituents of the part boundary.

Define the necessary tooling for the facing and contour turning of the sample part.

TOOLING-AUTOMATIC   AUTOMATIC tool change will issue the following prompts:
                                  (prompt)   (reply)
                                  TOOL No.:   1 ↵
                                  OFFSET No.:   1 ↵
                                  X-GA LENGTH: 0 ↵   For actual values
                                  Z-GA LENGTH: 0 ↵   use offset reg.
                                  TOOL NOSE RAD.:   .0312↵

Cutting data can be provided as:

OPTIONS-MACH. FUNCTIONS-FEED
FPR
(prompt)   (reply)
ENTER Feed per Revolution: .01 ↵
SPINDLE-CSS-FORWARD   This will generate MO3 in the tape file
(prompt)   (reply)
ENTER Feet per second: 500 ↵
SPINDLE-RANGE
(prompt)   (reply)
ENTER Range No.:   2 ↵ (High speed range)

SPINDLE-MAX.RPM
(prompt)   (reply)
ENTER Max. RPM   : 3000 ↻ (This will limit the maximum RPM in this process)
RETURN   From the fixed menu

The tool tip may at this point move to the start position for turning, as:

POSITION-TURN-LOCATION
DONE

(prompt)   (reply)
ENTER START POSITION COORDINATES [X,Z] : 5,6 ↻

It should be noticed that the tool point will be marked with a small circle.

(c.) The facing operation can best be done by developing a face boundary as:

-Change Color
GEOMETRY-BOUNDARY-FACE
(prompt)   (reply)
ENTER FACE LINE: .25 ↻   L10 will be displayed
DIA
(prompt)   (reply)
ENTER MIN. DIA.:   0 ↻   L11 will be displayed
FACE
(prompt)   (reply)
ENTER FACE NEAR PART: 0 ↻   L12 will be displayed
ELEMENT   Pick L7   L13 will be displayed
TERMINATE   This will display the new boundary in the new color, with
                                    L12 between L11 and L13 between L12 and L10.

To machine the face, select the following new items and answer the issued prompts as:

MACHIN-FACE IN   The facing tool will cut inward, toward the centerline.
ROUGHING   This will issue the prompts below:
(prompt)   (reply)
MIN. STOCK ON DIAs :   0 ↻   (If a finish operation is
MAX. STOCK ON DIAs :   0 ↻   required, allowance for it
MIN. STOCK ON FACE :   0 ↻   must be made here).
MAX. STOCK ON FACE :   0 ↻
MAX. DEPTH OF CUT :   .05 ↻
PICK PART BOUNDARY :   Pick the part boundary as described above.

The system will now display the tool path as shown in Fig. 12.17, Tool Path. This figure is a composite of the successively displayed tool paths.

(d.) Rough/finish turn contour.

```
-Change color.
MACHIN-TURN-ROUGH
(prompt)                                    (reply)
MIN. STOCK ON DIAs :                        .010 ↵
MAX. STOCK ON DIAs :                        .020 ↵
MIN. STOCK ON FACE :                        .010 ↵
MAX. STOCK ON FACE :                        .020 ↵
MAX. DEPTH OF CUT :                         .015 ↵
```

Other options, like EQUAL DEPTH, EQUAL RANGE or No. OF PASSES can also be chosen.

The preceding facing used part boundary, identification and turning. The programmer can, however, specify additional features or restrictions, such as:

OPTIONS-MACH-FUNCTION-CUTTERCOMP

This allows the programmer to use cutter compensation left or right.

```
RIGHT
(prompt)                                    (reply)
ENTER TOOL BANK No.:                        1 ↵
This will allow the operator to adjust for cutter wear.
PARAMS-CLEARANCE
```

```
(prompt)                                                                     (reply)
ENTER TOOL POINT CLEARANCE OVER PART IN RETURN CYCLE:                         .1 ↵
```

```
RETURN                              From the fixed menu
```

```
(prompt)                    (reply)
IDENTIFY PART               PICK the part boundary contour.
BOUNDARY:
```

The system will display the tool paths as shown by Figure 12.17.

Finish turn:

```
MACHIN-TURN-PROFILE-BOUNDARY-TOOL RIGHT
(prompt)                    (reply)
CUT IN DIRECTION            Yes ↵
DEFINED:
IDENTIFY PART               The programmer may PICK any element of the parth boundary
BOUNDARY:                       (L1, L2, . . up to L6).
```

It should be noticed that the tool path will be identical to the part contour (without, at this stage, the groove or the thread). Since a tool change will be required for the next operation, the tool will return to the home position as:

POSITION-HOME-X-FIRST   Notice the grooving tool's return path.
DONE   This will drive the tool turret to the home position.

　　　(e.) Turn the groove.

TOOLING-AUTO
(prompt)   (reply)
ENTER TOOL No. :   2 ↩
ENTER OFFSET No. :   2 ↩
ENTER X-GA LENGTH :   0 ↩
ENTER Z-GA LENGTH :   0 ↩
ENTER TOOL NOSE RAD :   0 ↩
WIDTH
(prompt)   (reply)
ENTER GROOVING WIDTH :   .125 ↩
OPTIONS-MACH FUNCT-SPINDLE-RANGE

(prompt)   (reply)
ENTER SPINDLE RANGE No. :   1 ↩
FEED-FPR
(prompt)   (reply)
ENTER FEED PER REVOLUTION :   .002 ↩
SPINDLE-CSS-FORWARD
(prompt)   (reply)
ENTER CONSTANT SURF. SPEED :   200 ↩
RETURN   From fixed menu
DONE

　　　For rapid approach before cutting :

POSITION-TURN-LOCATION-CURSOR POSN.
(prompt)   (reply)
PICK THE START POSITION FOR THE GROOVE TOOL :   PICK the approximate location of Q.
                                                 This point may be closer to the
                                                 part.

　　　For cutting:

POSITION-TURN-2SURFACES-TAN OFFSET
(prompt)   (reply)
IDENTIFY Z-OFFSET LINE :   PICK line LT
ENTER OFFSET AMOUNT :   .1 ↩
IDENTIFY X-OFFSET LINE :   PICK line L3
ENTER OFFSET AMOUNT :   .1 ↩

The tool point will rapid to .1 left to LT and .1 above L3, then proceed with the actual turning as:

MACHIN-GROOVE-O DIA	Outside Diameter
(prompt)	(reply)
IDENTIFY TOP DIA. OF GROOVE :	PICK L3
IDENTIFY BOTTOM DIA. OF GROOVE :	PICK LB
IDENTIFY LINE OF GROOVE FINISH OR ENTER Z-VALUE :	PICK LP, or enter the width of the groove and Return ↩.

ENTER OFFSET REGISTER No. : 3 ↩

The system will now display the tool path as shown by Figure 12.17.

Return tool to the home position as:

POSITION - HOME - X-FIRST DONE

RETURN of fixed menu. This will display the tool path on its return.

(f.) Turn thread.

TOOLING - AUTO

(prompt)	(reply)
ENTER TOOL No. :	3 ↩
ENTER OFFSET No. :	3 ↩
ENTER X-GA LENGTH :	0 ↩
ENTER Z-GA LENGTH :	0 ↩
ENTER TOOL NOSE RAD. :	0 ↩

OPTIONS-MACH FUNCTIONS-SPINDLE-RANGE

(prompt)	(reply)
ENTER RANGE :	1 ↩

SPINDLE - RPM

(prompt)	(reply)
ENTER RPM :	200 ↩

FEED

(prompt)	(reply)
ENTER FEED :	(1/20) ↩

DONE

NOTE: At this point it would be advisable to blank the tool paths in order to clear the graphics, as:

```
LAYERS - ON/OFF - NAME/NUM (Name or Number), or
LAYERS - ON/OFF - TOOL PATHS - ALL OFF
```

Move tool point close to thread as:

```
POSITION - TURN - LOCATION - CURSOR POSN
```

Digitize the approximate location of "S," in Figure 12.16.

```
MACHIN - THREAD - STD-OD - UNIFORM DIA
```

(prompt)	(reply)
ENTER THREAD DIA. :	1 ↵

```
12 UN
```

(prompt)	(reply)	
ENTER No. OF PASSES :	4 ↵	
PICK LINE OF Z-ENTRY :	PICK L12	
ENTER APPROX. DISTANCE :	.50 ↵	
PICK LINE AT END OF Z-VALUE :	PICK LT	
ENTER RELIEF DISTANCE :	.1 ↵	(for deceleration)

The system will again display the tool path. The programmer will now move the tool to home position, as:

```
POSITION - HOME - X-FIRST - DONE
END - END of fixed menu
DONE
```

In order to display all the tool paths,

```
MECH - MFG - NC-PROGRAM - OLD PROMPT
DONE - DONE
```

Pick from list. All the tool paths will be reactivated in the sequence of the process.

(prompt)	(reply)
SAVE THE PART PROGRAM ON:	TURN1 ↵ ↵ ↵

To check the source, the programmer can display it, print a hard copy before the tape file is prepared, or type a tape file as:

UTILITIES-HOST COMMAND

(prompt)   (reply)
ENTER VMS COMMAND :   TYPETURN1.TAP⤶

DONE
TERMINAL

This will display the program on the screen. It will be noticed that the tape will
have several "@END" lines in it. These must be edited out before the tape is
punched, as:

EXIT   From main menu

(prompt)   (reply)
$ PUNCH/NTAPE/LIST = TURN1G TURN1.TAP

This will edit the "@END" out of the program.
    It should be noted that this program is far easier to work with on the screen,
with its menu-driven structure, than the description presents it on paper. It can
also be seen that the programmer does not have to know the exact definitions of
either geometries or motion commands. The prompts are relatively easy to follow.

12.10.3 BRAVO3 Output Program

```
$APPL SRC 00 NCG 19890921170055 0 NCG 19890921170055 0 0 37218
$DUA1:[STAFF.PUSZTAI.EQUINOX]PROFILE.DDD
$APPL SRC 00 3DC 19890920150421 0 NCG 19890920150424 0 0 41452
MACHIN,9400,57090
INDENT, THIS IS A TEST OF PROGRAM IN THE BOOK
INIT,INCH/IN,INCH/OUT
SETUP,5X,4Z,CMOD(6,1,1),MOD(3,1),ZEROS3
LAYER,0,0,0,-5,-1,2,8,-6,2,3,3,1,10,9,-7,-4,4
$\ \ \START*EXTERNAL*GEOMETRY/ / /        16
DRAW,FRAME,PT(ZB-5.333333,XB-1.33333)   &
  ,PT(ZB1.333333,XB3.083333)
DRAW,PEN1              $ >>> RED <<<
DLN11,SLAYER2,PT(ZB0,XB0),PT(ZB-4,XB0)
DLN1,SLAYER2,PT(ZB0,XB.442265),PT(ZB0,XB0)
DLN3,SLAYER2,PT(ZB-.84,XB.5),PT(ZB-.1,XB.5)
DLN14,SLAYER2,PT(ZB-.84,XB.5),PT(ZB-.84,XB.42)   $LT
DLN15,SLAYER2,PT(ZB-1,XB.5),PT(ZB-1,XB.42)   $  LP
DLN16,SLAYER2,PT(ZB-.84,XB.42),PT(ZB-1,XB.42)   $LB
DLN5,SLAYER2,PT(ZB-2.45,XB1.5),PT(ZB-2.875,XB1.5)
DLN4,SLAYER2,PT(ZB-1,XB.5),PT(ZB-2.11226,XB1.05613)
DCIR2,SLAYER2,PT(ZB-2.250203,XB1.290994),PT(ZB-2.216025,XB1.153137),PT(ZB-2.11226,XB1.05613)
```

```
DCIR1,SLAYER2,PT(ZB-2.25,XB1.3),PT(ZB-2.308579,XB1.441421),PT(ZB-2.45,XB1.5)
DLN2,SLAYER2,PT(ZB-.1,XB.5),PT(ZB0,XB.442265)
DLN6,SLAYER2,PT(ZB-2.875,XB1.5),PT(ZB-3,XB1.75)
DLN7,SLAYER2,PT(ZB-3,XB1.75),PT(ZB-4,XB1.75)
$\\\END*EXTERNAL*GEOMETRY///
DPT26216,XB,ZB
$\\\NCG*TOOLBUF///
DRAW,PEN2
DLN12,SLAYER6,ZB0
DLN18,SLAYER6,D-.05
DLN13,SLAYER6,D3.75
TRMLN12,PT(-.0061ZB,1.1785XB),SLAYER6,F(LN13),S(LN18),TRIMTO
TRMLN18,PT(.2167ZB,-.0133XB),SLAYER6,F(LN12),TRIMTO
TRMLN13,PT(.5175ZB,1.897XB),SLAYER6,F(LN12),TRIMTO
DLN10,SLAYER6,LN12/ZS-.25
TRMLN13,PT(.122ZB,1.8803XB),SLAYER6,F(LN10),TRIMTO
TRMLN18,PT(.1387ZB,-.0412XB),SLAYER6,F(LN10),TRIMTO
ATCHG,TOOL1,OFFSET1,GLX0,GLZ0,.03125TLR,TLAYER7,IPR.01,FPM500,RANGE2,MAXRPM30000
MOVEC,PT(ZB5,D4),CSS2/ON
DRAW,PEN6
MOVEC,PT(LN10,NRPT(.2557ZB,3.6715D),Z.1,X.1)
DPB2,SLAYER6,LN13,NRPT(.122ZB,3.7829D)
DPB2,LN12,NRPT(.0162ZB,3.5156D)
DPB2,LN18,NRPT(.1332ZB,.0067D)
DPB2,LN10,NRPT(.2501ZB,.2406D)
DPB2,NOMORE
CUT,PARX,BARSTK,XSTK(0)/0,ZSTK(0)/0,MAXDP.05,PB2
DRAW,PEN5
DPB3,SLAYER6,LN1,NRPT(.0051ZB,.4634D)
DPB3,LN2,NRPT(-.0451ZB,.9201D)
DPB3,LN3,NRPT(-.2511ZB,1.0538D)
DPB3,LN4,NRPT(-1.2203ZB,1.2432D)
DPB3,CIR2,NRPT(-2.2005ZB,2.3682D)
DPB3,CIR1,NRPT(-2.3676ZB,2.9363D)
DPB3,LN5,NRPT(-2.6795ZB,3.0477D)
DPB3,LN6,NRPT(-2.9691ZB,3.1925D)
DPB3,LN7,NRPT(-3.1696ZB,3.4933D)
DPB3,NOMORE
DRAW,PEN1
CUT,BARSTK,XSTK(.01)/.02,ZSTK(.01)/.02,MAXDP.1,PB3
DRAW,PEN2
CUT,PB3/CR
HOME,XF
ATCHG,TOOL2,OFFSET2,GLX0,GLZ0,.125TLW,0TLCR,TLAYER9,IPR.002,FPM002,RANGE1
MOVEC,OFFLN7/XL.1,OFFLN15/ZL.05,CSS2/ON
MOVEC,OFFLN14/ZS.125,OFFLN3/XL.25
GRV,T(LN3),B(LN16),S(LN14),F(LN15)
HOME,XF
```

```
ATCHG,TOOL3,OFFSET3,GLX0,GLZ0,0TLR,TLAYER10,IPR(1/12),RPM200,RANGE1
MOVEC,LN(.2724ZB),LN(2.0752XB)
DRAW,PEN6
MOVEC,OFFLN1/ZL.5,OFFLN3/XL.1
THRD,CX,MAD1.,12UN,PASS6,S(LN1,CLDIST.2),F(ZB-.84,CLDIST.1)
HOME,XF
END
```

12.10.4 BRAVO3 Output Tape File

```
MORI FAN 49-I      THIS IS A TEST OF PROGRAM IN THE BOOK
21-SEP-89   17:01
%
N10G20
N20G50X10.Z4.
N30G00T0100
N40M42
N50G50S3000
N60G96S0500M03
N70G04U1.472
N80X4.0626Z5.T0101
N90X4.0126Z.35
N100X3.9126Z.2313
N110G99G01X.0F.01
N120Z.2813
N130X.1126Z.3313
N140G00X3.9126
N150Z.1813
N160G01X.0
N170Z.2313
N180X.1126Z.2813
N190G00X3.9126
N200Z.1313
N210G01X.0
N220Z.1813
N230X.1126Z.2313
N240G00X3.9126
N250Z.0813
N260G01X.0
N270Z.1313
N280X.1126Z.1813
N290G00X3.9126
N300Z.0313
N310G01X.0
N320Z.0813
N330G00X4.0126Z.35
```

N340X3.3626Z.0813
N350G01Z-2.9195
N360X3.5626Z-2.9695
N370G00Z.0813
N380X3.1626
N390G01Z-2.8695
N400X3.3626Z-2.9195
N410G00Z.0813
N420X2.9626
N430G01Z-2.2908
N440G03X3.0826Z-2.45I-.1813K-.1592
N450G01Z-2.8495
N460X3.1626Z-2.8695
N470X3.2626Z-2.8195
N480G00Z.0813
N490X2.7626
N500G01Z-2.2229
N510G03X2.9626Z-2.2908I-.0813K-.2272
N520G00Z.0813
N530X2.5626
N540G01Z-2.2088
N550X2.6
N560G03X2.7626Z-2.2229K-.2413
N570G00Z.0813
N580X2.3626
N590G01Z-2.1845
N600G02X2.532Z-2.2088I.0985K.1841
N610G01X2.5626
N620X2.6626Z-2.1588
N630G00Z.0813
N640X2.1626
N650G01Z-2.0703
N660X2.186Z-2.0938
N670G02X2.3626Z-2.1845I.1867K.0934
N680G00Z.0813
N690X1.9626
N700G01Z-1.8703
N710X2.1626Z-2.0703
N720G00Z.0813
N730X1.7626
N740G01Z-1.6703
N750X1.9626Z-1.8703
N760G00Z.0813
N770X1.5626
N780G01Z-1.4703
N790X1.7626Z-1.6703
N800G00Z.0813
N810X1.3626

```
N820G01Z-1.2703
N830X1.5626Z-1.4703
N840G00Z.0813
N850X1.1626
N860G01Z-1.0703
N870X1.3626Z-1.2703
N880G00Z.0813
N890X.9626
N900G01Z.015
N910X1.0714Z-.0794
N920G03X1.0826Z-.1I-.0357K-.0206
N930G01Z-.9903
N940X1.1626Z-1.0703
N950X1.2626Z-1.0203
N960G00Z.0813
N970X.886
N980G01X.9626Z.015
N990X1.0626Z.065
N1000G00X4.0126Z.35
N1010X.8846Z.0813
N1020G01Z.0625
N1030X1.0542Z-.0844
N1040G03X1.0626Z-.1I-.0271K-.0156
N1050G01Z-.9926
N1060X2.1682Z-2.0983
N1070G02X2.5596Z-2.2192I.1957K.0978
N1080X2.5796Z-2.219K.2188
N1090G03X2.6Z-2.2188I.0102K-.231
N1100X3.0626Z-2.45K-.2312
N1110G01Z-2.8557
N1120X3.5626Z-2.9807
N1130G00X4.0126Z.35
N1140X10.T0100
N1150Z4.
N1160G20
N1170G50X10.Z4.
N1180G00T0200
N1190M41
N1200G50S1142
N1210G96X3.70Z-.95S0200T0202M03
N1220X1.50Z-.965
N1230X1.10
N1240G99601X.84F.002
N1250G00X1.10
N1260Z-1.
N1270G01X.84
N1280G00X1.10Z-.9825
N1290X10.T0200
```

```
N1300Z4.
N1310G20
N1320G50X10.Z4.
N1330G00T0300
N1340M41
N1350G97X4.1504Z.2724S0200T0303M03
N1360X1.2Z.5
N1370X1.10Z.2
N1380G99G01X.9830F.0833
N1390G32Z-.94E.083333
N1400G00X1.10
N1410Z.2
N1420G01X.966F.0833
N1430G32Z-.94E.083333
N1440G00X1.10
N1450Z.2
N1460G01X.9490F.0833
N1470G32Z-.94E.083333
N1480G00X1.10
N1490Z.2
N1500G01X.9318F.0833
N1510G32Z-.94E.083333
N1520G00X1.10
N1530Z.2
N1540G01X.9148F.0833
N1550G32Z-.94E.083333
N1560G00X1.10
N1570Z.2
N1580G01X.8978F.0833
N1590G32Z-.94E.083333
N1600G00X1.10
N1610Z.2
N1620X10.T0300
N1630Z4.
N1640M30
%
```

12.11 IN-PROCESS GAGING

A relatively new area, this field is also known under the names of "probing," or "on-the-machine" measuring systems. As "machine-to-machine" communications become less of a problem, the programming and application of probes develop into one of the most important additions to the world of CNC.

Probing is normally used for:

- Locating a part on the machine table,

- Measuring the stock left on for machining, measuring castings or forgings, and
- Measuring the accuracy of the finished part.

There are several makes on the market today which compete in assisting the improvement of economics and accuracy of machined or fabricated components. Most of these systems are designed and used for on the machine measuring, there-

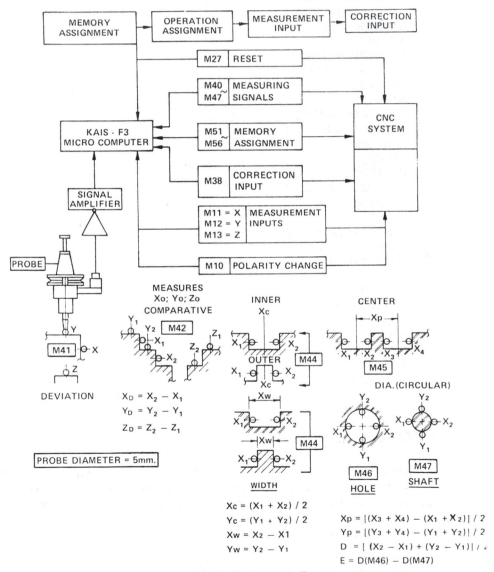

Figure 12.18 Probing logic diagram.

fore this discussion will be limited to an "Automatic Measurement Compensation" (AMC) system.

The probe under discussion is a KAIS-3A system, used in conjunction with an OKK-410 Machining Center (Fanuc 6MB Controller) or a Mori-Seiki SL3 Turning Center (Fanuc 6TB Controller). The probe is mounted in the spindle or in the turret at a specific position (by the automatic tool change command location), and when the probe is active, the spindle or the turret are automatically locked in that position.

The probe is also connected to the CNC electronically, to enable appropriate exchange of the programmed and the measured data information. The *measured deviation* can then be programmed to replace or update a desired tool offset register to maintain a programmed machining tolerance. The system also allows the *automatic recording* of the measured data.

The logic diagram illustrated in Figure 12.18 describes the operation of the KAIS probe. The measuring ranges are ± 2 mm on the X- and Y-axes, and ± 0.5 mm on Z. The probe will allow an overtravel of ± 15 mm on X and Y, and ± 4.5 mm on Z.

The probe logic diagram contains some additional M-codes. These are available as options at the time of the equipment purchase. The sample programs below will also show the need for additional G-codes, required for the AMC system.

Memory Assignments.   The programmer must identify the locations for measuring in memory, as well as assign locations for storage of the measurements and corrections. When the probe circuit is activated, the reserved memory is automatically assigned to M51. (Single probings do not require assignment by programming.)

The programming of the AMC system will be illustrated below for the following cases:

1. Deviation probing (M41)
2. Comparative probing (M42)
3. Center probing (M44)

12.11.1 Deviation Probing or Measurement

Task.   A part program is required to locate the toleranced dimensions of the part below. See Figure 12.19.

Assumptions

- The location of the G55 was assigned either by using Manual Data Input or by programming (G10 P4 L2 X60.00 Y20.00).
- Probe diameter is 5 mm.
- The tolerances were set in the KAIS computer prior to the running of the program.

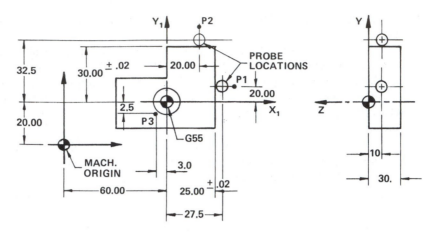

Figure 12.19 Deviation probing M41.

Solution

O100	*Program Number*
N05 G21 G40 G80	Metric input, cancel cutter diameter compensation and canned cycles.
N10 T11	Index to tool position 11, no offset.
M06 T12	Change tool, then index to tool position 12.
G28 G91 X0. Y0. Z0.	Return to reference point, incremental programming.
M27	Reset KAIS.
M38	Reset CNC measuring register to zero.
M41	Activate deviation measurement routine.
G00 G90 G55 X30. Y20. Z100.	Rapid to clear part for first measurement. Work offset No. 2. G55 was placed into the CNC by MDI prior to this operation.
G01 Z-10. F2000	Feed 10 mm down to measuring level.
G01 G9 X27.5 F1000	Exact stop check of the 27.5 mm dimension. Code G9 is auto deceleration.
M11	Place the measured X_2 dimension into KAIS.
G00 X30.	Rapid to clear the part.
G00 Y35.	Rapid to second measurement location in Y.
G00 X20.	Rapid to second measurement location in X.
G01 G9 Y32.5 F1000	Exact stop check of the 32.5 mm dimension.
M11	Place the measured X_1 dimension into KAIS.
G00 Y35.0 Z10.0	Rapid to clear part.
G28 G91 Z0.	Rapid to home in the Z-axis.
G28 G91 X0. Y0.	Rapid to home in the X- and Y-axes.
N20 M06	Change tool.
M38	Transfer the measured values from KAIS to CNC.
#2502 = #2502 + #2500	Update the X work offset.
#2602 = #2602 + #2600	Update the Y work offset. These values can be checked in these registers.

#2500 = 0	Reset the X register to 0.
#2600 = 0	Reset the Y register to 0.
G00 G90 G55 X0. Y0.	This confirmation measurement will show ± the deviation, as a result of the above probing.
M30	End of program or return to main program.
%	

The confirmation block will *relocate* the spindle centerline onto the G55 work offset location ± the measured deviation. For instance, if the X27.5 was measured as 30.5, and the Y32.5 was measured as 35.00, the G55 (previously programmed work offset) location will be moved to the location of P3 in order to maintain the required dimensions of 25.00 ± 0.02 and 30.00 ± 0.02.

When the measured dimensions are precisely 25.00 and 30.00, the controller measurements will be 0.0 mm. By this method the dimensional errors can be detected and compensated for.

If the measured dimensions are within tolerances (high +0.02 and low −0.02, preset in the KAIS computer under the M41 program function), the system will carry on, i.e., the program can proceed to machine the block. If on the other hand the measured dimension is outside the above preset tolerances, the system will display No Good (NG) and will turn the Machine Hold circuit on, therefore requiring the operator to reset.

12.11.2 Comparative Probing or Measurement.

Task.   Write a part program to find the X-distance between the two measured faces of the part below (dimension 15 ± 0.02). See Figure 12.20.

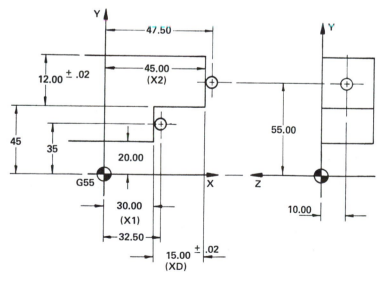

Figure 12.20   Comparative measurement M42.

Assumptions.

- Location of the work offset, G55, was set prior to the running of this program.
- Probe diameter = 5.00 mm.
- The program will perform the X-axis directional comparison only (X_D) as programmed below.

Solution.

O1001	*Program Number*
N05 G21 G40 G80	Metric input, cancel cutter diameter compensation and canned cycles.
N10 T11	Index to Tool Position 11
M06 T12	Change tool, then index to tool position No. 12
G28 G91 X0. Y0. Z0.	Return to reference point, incremental programming.
M27	Reset KAIS.
M38	Reset the CNC measuring register to zero.
M42	Activate comparative measurement assignment.
G00 G90 G55 X50. Y55. Z100.	Rapid to clear part for first measurement; work offset No. 2. G55 was placed into the CNC by MDI prior to this operation.
G01 Z-10. F2000	Feed 10 mm down to measuring level.
G01 G9 X47.5 F1000	Exact stop check of the 47.5 mm dimension.
M11	Place the measured X_2 dimension into KAIS.
G00 X55.0	Rapid to clear the part.
G00 Y35.0	Rapid to second measurement location in Y.
G00 X35.0	Rapid to second measurement location in X.
G01 G9 X32.5 F1000	Exact stop check of the X32.5 dimension.
M11	Place the measured X_1 dimension into KAIS.
G00 X35.0 Z10.0	Rapid to clear part.
G28 G91 Z0.0	Rapid to home in the Z-axis.
G28 G91 X0.0 Y0.0	Rapid to home in the X- and Y-axes.
N20 M06	Change tool.
M38	Transfer the measured values from KAIS to CNC.
#2502 = #2502 + #2500	Update the X work offset.
#2602 = #2602 + #2600	Update the Y work offset. These values can be checked in these registers.
#2500 = 0	Reset the X register to zero.
#2600 = 0	Reset the Y register to zero.
G00 G90 G55 X0. Y0.	This confirmation measurement will show ± the deviation as a result of the above probing.
M30	

The limits and parameters must be set prior to running of this program, then the X_D will be calculated to 1 micrometre accuracy in terms of X_2 and X_1. The same program can easily be expanded to probe Y or Z dimensions as well.

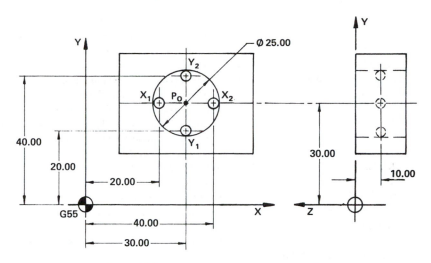

Figure 12.21   Center location measurement M44.

12.11.3 Center Measurement

Task. Write a part program to find the center point P_0 of the component illustrated below. See Figure 12.21.

Assumptions.

- Location of the work offset, G55, was set prior to the running of the program.
- Probe ball diameter = 5 mm.

Solution.

O1002	*Program Number*
N05 G21 G40 G80	Metric input. Cancel cutter diameter compensation and canned cycles.
N10 T11	Index to tool position 11.
M06 T12	Change tool then index to tool position 12.
G28 G91 X0. Y0. Z0.	Return to reference point, incremental programming.
M27	Reset KAIS.
M38	Reset CNC measuring register to zero.
M44	Activate center measurement assignment.
G00 G90 G55 X30. Y30. Z100.	Rapid to center of hole to be measured. Work offset No. 2. G55 was placed into the CNC by MDI prior to this operation.
G01 Z-10. F2000	Feed 10 mm down into hole.
G01 G9 X20. F1000	Exact stop check of the 20.0 mm dimension.
M11	Place the measured X_2 dimension onto KAIS.
G01 G9 X40.0 F1000	Exact stop check the 40.0 mm dimension.

M11	Place the measured X_1 dimension into KAIS. Note that the KAIS will perform the required calculations and will display the X_w and X_c dimensions on the KAIS screen.
G00 X30.0	Rapid to the center of the hole.
G01 G9 Y20.0 F1000	Exact stop check the Y 20.0 mm dimension.
M12	Place this measurement into KAIS.
G01 G9 Y40.0 F1000	Exact stop check the Y 40.0 mm dimension.
M12	Place this measurement into KAIS. Note that the KAIS will perform the required calculations and will display the Y_w and Y_c dimensions on the KAIS screen.
G00 Y30.0	Rapid to the center of hole.
Z100.	Move probe to clear part.
G28 G91 Z0.0	Rapid to home in the Z-axis.
G28 G91 X0.0 Y0.0	Rapid to home in the X- and Z-axes.
N20 M06	Change tool.
M38	Transfer the measured values from KAIS to CNC.
#2502 = #2502 + #2500	Update the X work offset.
#2602 = #2602 + #2600	Update the Y work offset. These values can be checked in these registers.
#2500 = 0	Reset the X register to zero.
#2600 = 0	Reset the Y register to zero.
G00 G90 G55 X0. Y0.	This confirmation measurement will show $\pm$ the deviation as a result of the above probing.
M30	
%	

Based on the above measurements, the actual center point P_0, or the preprogrammed location of G55, will be relocated to the actual position. This method of probing can be used for locating the center point of an existing hole or shaft for purposes such as finish machining the hole, drilling a circular pattern around the hole, etc. One must realize that the outside contour cannot be located for machining purposes with this method. Actual toleranced sizes of holes or shafts can also be probed using the M46 or M47 option.

This technique can be used, in addition to probing, for in-process gaging. The length and complexity of that program are beyond the scope of this text, and its user friendliness requires improvement as well. Some highlights are given below.

The program may use variables, similar to the ones in previous programs. In the G65 block, the program may assign a whole schedule of gaging procedures, such as G65 P100 I1 K5 I5 K2 I10 . . . etc., where

> I1 : Instruction to check each piece.
> K5 : Previous instruction applies to the first five pieces.
> I5 : Check fifth piece specifically.

K2 : Check (that fifth piece) two times.
I10 : Check only every tenth piece from now on. etc.

It should be noted that in this type of program, *I* and *K* work in pairs, with "*K*" complementing the instruction provided by "*I*".

The G31 (skip cutting) is used for measuring as:

G31 X#2 F0.8

The X dimension is measured, and its value stored into register No. 2.

#20 − ABS [OPEN 5061]

Variable #20 is assigned the value measured in the previous line. This value is stored into register 5061.

G10 P11 X#20

Transfer the contents of previously assigned variable into offset register No. 11, in compartment X.

#21 = #2014 − #20

This is a way to find the difference between tool offset No. 2014 and variable No. 20, and then store this value into another offset register, No. 12, using the instruction:

G10 P12 #21

Using this method, the programmer can continuously update the values of the tool offset registers, compensate for tool wear and control the programmed tolerances.

EXERCISES

12.11.1. Modify the comparative probing program to measure or probe the 12.00 and the 45 mm dimensions along the Y axis.

12.11.2. The following is the beginning of a probing part program for a comparative probing application:

O100
N10 T01
M06 T0300

G04 X100
G91 G28 X0.0 Y0.0 Z0.0

Make the necessary corrections for proper start-up.

12.11.3. Write down the formula and sketch the application for the probing of the center Y_P of two parallel grooves in the "Y" direction. Also, identify the proper M code for same.

12.11.4. Write down the formula and sketch the Y_C application for the probing of the centerline and the outer wall's thickness Y_W in the Y-direction Y_C; also identify the proper M-code for same.

Appendix A

TABLE A-1 MISCELLANEOUS FUNCTIONS (M-CODES)

M-Code	Mill/Turn	Electrical Discharge Machining-EDM
M00	Program stop	Program stop
M01	Optional stop	Optional stop
M02	End of program and tape rewind	End of program and reset
M03	Spindle start CW	Electrode rotation ON
M04	Spindle start CCW	—
M05	Spindle stop	Electrode rotation OFF
M06	Tool change	—
M08	Coolant ON	—
M09	Coolant OFF	—
M10	Polarity change (probing)	—
M11	Read X-measurement (probing)	—
M12	Read Y-measurement (probing)	—
M13	Read Z-measurement (probing)	—
M19	Spindle orient/stop	—
M20	—	Wire insert
M21	Mirror image X	Wire cut
M22	Mirror image Y	Test insert
M23	Mirror image OFF	Fixed position return
M27	Reset register (probing)	—

(Continued)

TABLE A-1 (*Continued*)

M-Code	Mill/Turn	Electrical Discharge Machining-EDM
M30	End of program and memory rewind	Program end, rewind, and reset
M38	Measurement correction to CNC	—
M41	Low range (mill)	—
	Deviation measurement (probing)	—
M42	High range (mill)	—
	Comparative measurement (probing)	—
M44	Center location measurement (probing)	—
M45	Multisurface center locations (probing)	—
M46	Hole center (probing)	Initial hole compensation 1
M47	Shaft center	Initial hole compensation 2
M48	Override cancel OFF (mill)	Initial hole compensation 3
M49	Override cancel ON (mill)	Initial hole compensation 4
M60	—	External signal output 1 ON
M61	—	External signal output 1 OFF
M62	—	External signal output 2 ON
M63	—	External signal output 2 OFF
M64	—	External signal output 3 ON
M65	—	External signal output 3 OFF
M66	—	External signal output 4 ON
M67	—	External signal output 4 OFF
M68	—	External signal output 5 ON
M69	—	External signal output 5 OFF
M70	—	Initial hole mode ON
M71	—	Initial hole mode OFF
M74	—	Dielectric fluid preparation ON
M75	—	Dielectric fluid preparation OFF
M76	—	Dielectric fluid circulation ON
M77	—	Dielectric fluid circulation OFF
M78	—	Quick filling up
M80	—	Dielectric fluid ON
M81	—	Dielectric fluid OFF
M82	—	Wire feed ON
M83	—	Wire feed OFF
M84	—	Machining power supply ON
M85	—	Machining power supply OFF
M88	—	Decrease dielectric fluid flow
M89	—	Increase dielectric fluid flow
M90	—	Adaptive control feed ON
M91	—	Adaptive control feed OFF
M93	—	All stop
M95	—	Automatic machining ON
M96	—	Automatic machining OFF
M98	Go to subroutine	—
M99	Return from subroutine	—

TABLE A-2   PREPARATORY FUNCTIONS (G-CODES)

G-Code	Machining Centers
G00	Positioning (rapid traverse)
G01	Linear interpolation (cutting feed)
G02	Circular/helical interpolation (clockwise)
G03	Circular/helical interpolation (counterclockwise)
G04	Dwell cycle
G09	Exact stop, deceleration
G10	Offset value setting by program
G14	—
G15	Polar coordinates command cancel
G16	Polar coordinates command
G17	X-Y plane selection
G18	Z-X plane selection
G19	Y-Z plane selection
G20	Inch data input (G70 on some systems)
G21	Metric data input (G71 on some systems)
G22	Safety zone programming
G23	Programmed crossing through safety zone
G25	—
G27	Reference point return check
G28	Return to reference point
G29	Return from reference point
G30	Return to 2nd, 3rd, or 4th reference point
G31	Skip function
G32	—
G33	Thread cutting
G37	Tool length automatic measurement
G38	Cutter diameter compensation vector change
G39	Cutter diameter compensation corner rounding
G40	Cutter diameter compensation cancel
G41	Cutter diameter compensation left
G42	Cutter diameter compensation right
G43	Tool length compensation + direction
G44	Tool length compensation − direction
G45	Tool offset increase
G46	Tool offset decrease
G47	Tool offset double increase
G48	Tool offset double decrease
G49	Tool length compensation cancel
G50	Scaling cancel
G51	Scaling
G50.1	Programmable mirror image cancel
G51.1	Programmable mirror image
G52	Local coordinate system setting
G53	Machine coordinate system selection
G54	Work coordinate system 1 selection

(Continued)

TABLE A-2 (*Continued*)

G-Code	Machining Centers
G55	Work coordinate system 2 selection
G56	Work coordinate system 3 selection
G57	Work coordinate system 4 selection
G58	Work coordinate system 5 selection
G59	Work coordinate system 6 selection
G60	Single direction positioning
G61	Exact stop mode
G62	Automatic corner override mode
G63	Tapping mode
G64	Cutting mode
G65	Transfer to subroutine (macro call)
G66	—
G68	Coordinate system rotation
G69	Coordinate system rotation cancel
G70	—
G71	—
G73	Peck-drilling canned cycle
G74	Counter tapping canned cycle
G76	Fine boring canned cycle
G80	Canned cycle cancel
G81	Drilling/spot drilling canned cycle
G82	Drilling/counter boring canned cycle
G83	Peck drilling canned cycle
G84	Tapping canned cycle
G85	Boring canned cycle
G86	Boring canned cycle
G87	Back boring canned cycle
G88	Boring canned cycle
G89	Boring canned cycle
G90	Absolute data input
G91	Incremental data input
G92	Work coordinates change—absolute zero point
G93	—
G94	Feed per minute
G95	Feed per revolution
G96	Constant surface speed control
G97	Constant surface speed control cancel
G98	Canned cycle—return to Initial level
G99	Canned cycle—return to R-level

TABLE A-3 PREPARATORY FUNCTIONS (G-CODES)

Std. G-Code	Spec. G-Code	Turning Centers
G00	G00	Positioning (rapid traverse)
G01	G01	Linear interpolation (cutting feed)
G02	G02	Circular interpolation (clockwise)
G03	G03	Circular interpolation (counterclockwise)
G04	G04	Dwell cycle
G09		—
G10	G10	Offset value setting by program
G14		—
G15		—
G16		—
G17		—
G18		—
G19		—
G20	G70	Inch data input
G21	G71	Metric data input
G22	G22	Safety zone programming
G23	G23	Programmed crossing through safety zone
G25		—
G27	G27	Reference point return check
G28	G28	Return to reference point
G29	G29	Return from reference point
G30	G30	Return to 2nd reference point
G31	G31	Skip cutting
G32		—
G34	G34	Variable lead thread cutting
G36	G36	Automatic tool compensation X axis
G37	G37	Automatic tool compensation Z axis
G38		—
G39		—
G40	G40	Tool nose radius compensation cancel
G41	G41	Tool nose radius compensation left
G42	G42	Tool nose radius compensation right
G43		—
G44		—
G45		—
G46		—
G47		—
G48		—
G49		—
G50	G92	Programming of absolute zero point Maximum spindle speed setting
G51		—
G50.1		—
G51.1		—
G52		—

(Continued)

TABLE A-3 (*Continued*)

Std. G-Code	Spec. G-Code	Turning Centers
G53		—
G54		—
G55		—
G56		—
G57		—
G58		—
G59		—
G60		—
G61		—
G62		—
G63		—
G64		—
G65	G65	Transfer to subroutine (macro call)
G66	G66	Modal user macro call command
G68	G68	Mirror image for double turrets ON
G69	G69	Mirror image for double turrets OFF
G70	G72	Finishing cycle
G71	G73	Stock removal in turning
G72	G74	Stock removal in facing
G73	G75	Pattern repeating
G74	G76	Peck-drilling in Z axis
G75	G77	Grooving in X axis
G76	G78	Thread cutting cycle
G80		—
G81		—
G82		—
G83		—
G84		—
G85		—
G86		—
G87		—
G88		—
G89		—
G90	G20	Cutting cycle A
G91	—	—
G92	G21	Thread cutting cycle
G93	—	—
G94	G24	Cutting cycle B
G95	—	—
G96	G96	Constant surface speed control on
G97	G97	Constant surface speed control cancel
G98	G94	Feed per minute
G99	G95	Feed per revolution
—	G90	Absolute programming
—	G91	Incremental programming

TABLE A-4 PREPARATORY FUNCTIONS (G-CODES)

G-Code	Electrical Discharge Machining (EDM)
G00	Positioning (rapid traverse)
G01	Linear interpolation (cutting feed)
G02	Circular/helical interpolation clockwise
G03	Circular/helical interpolation counterclockwise
G04	Dwell cycle
G09	—
G10	—
G14	Current position reading
G15	Printout
G17	—
G18	—
G19	—
G20	—
G21	—
G22	Transfer to subroutine (macro call)—figure rotation
G23	Return from subroutine—figure rotation
G25	—
G27	Zero point collation
G28	Return to reference point
G29	Automatic edge positioning
G30	Automatic center positioning
G32	—
G39	—
G40	Wire diameter compensation cancel
G41	Wire diameter compensation left
G42	Wire diameter compensation right
G43	Program wire diameter compensation distance setting
G44	Program wire diameter compensation distance transmitting
G45	—
G46	Microjoint function
G47	—
G48	Return to 2nd, 3rd, or 4th reference point
G49	—
G50	—
G60	Dog-type zero point return
G61	—
G62	Programmable mirror image
G70	Inch data input
G71	Metric data input
G87	Upper/lower identical R
G88	Corner R-mode
G89	Intersection point calculation mode
G90	Absolute data input
G91	Incremental (relative) data input

(Continued)

TABLE A-4   *(Continued)*

G-Code	Electrical Discharge Machining (EDM)
G92	Work coordinates change–zero point preset
G93	—
G94	—
G95	—
G96	—
G97	X-Y axis switch
G98	—
G99	—
G101	$a = b$
G102	$a = b + c$
G103	$a = b - c$
G104	$a = b * c$
G105	$a = b/c$
G106	$a = \sqrt{b2 + c2}$
G107	$a = b * \sin c$
G108	$a = b * \cos c$
G109	$a = \tan^{-1} b/c$
G110	$a = \sqrt{b2 - c2}$
G200	Unconditional branch (go to . . .)
G201	Null condition branch (if zero, go to . . .)
G202	Negative condition branch (if negative, go to . . .)
G203	Bit test

NOTE: The preceding tables represent another example of the fact that standardization among the various manufacturers of controllers and machine-tool manufacturers who select or assign specific codes to specific functions is still a challenge.

If one examines a number of specific G codes in Tables A-2, A-3, and A-4, it will be seen that the same code may represent a number of different functions.

It is not the intent of the authors to state that controller standardization from the point of view of the end-user is an easy task. However, as the controllers become more and more capable of new functions, the proliferation of codes goes against the requirement of user friendliness. This creates confusion among the people whose goal is to become proficient in the use of this equipment, be they college students or industry programmers and operators.

The specific operator's and programmer's manual for one's equipment must be identified and followed at all times.

Index

A

Absolute Programming, 71
ACL, 62
Advanced CNCs, 56
Allen Bradley, 60
Analytical Geometry, 30
APT, 215, 216, 218, 220, 221
  Listing, 255
  Part Programming, 252
  Tape File, 257
ASCII Cutter Location, 62
ASCII Tape Code, 17
Assignment of Argument
  for variables, 181
Automatic Repeat Cycles, 124
Automation, 1

B

BCD, 5, 14, 15
BCL, 60

C